J ZUFELT

World Class
Maintenance
Management

Terry Wireman

World Class Maintenance Management

Industrial Press Inc.

Library of Congress Cataloging-in-Publication Data

Wireman, Terry.
World class maintenance management / Terry Wireman. — 1st ed.
192 p. 15.6 × 23.5 cm.
ISBN 0-8311-3025-3
1. Plant maintenance—Management. 2. Industrial equipment–
–Maintenance and repair. I. Title.
TS192.W58 1990
658.2′02—dc20 90-32673
CIP

ISBN 0-8311-3025-3

Industrial Press Inc.
200 Madison Avenue
New York, New York 10016-4078

First Edition
World Class Maintenance Management

Composition by David E. Seham Associates, Metuchen, New Jersey.
Printed and bound by Quinn Woodbine, Woodbine, New Jersey.

9 10 11 12 13 14 15

Preface

Maintenance—many times this word is viewed negatively. In almost all organizations, the maintenance function is viewed as a necessary, disaster-repairing function. However, little time or effort is spent on trying to control maintenance activities and costs.

While maintenance, in the past, has received little notice, most organizations are presently trying to control costs. Since maintenance costs are very high for most organizations, much attention is being turned to accountability for maintenance expenditures. As organizations audit their maintenance costs, they find a sizable amount of money spent with little management control.

Proper management controls must be applied to maintenance if costs are to be curbed. But to control maintenance successfully, proper management must be instituted. Again, many organizations have tried to use standard production- or facilities-oriented methods to control maintenance. This will not work.

Maintenance is a discipline of its own. It requires a different approach to manage successfully.

It is the purpose of *World Class Maintenance Management* to present an insight into what is required to manage maintenance. It is not a total answer to every maintenance management problem; however, it will provide a framework with options, allowing maintenance decision makers to select the most successful way for them to manage their maintenance.

Contents

Preface v

Introduction—Status of Maintenance in the United States 1

1 Analyzing Maintenance Management 12

2 Maintenance Organizations 32

3 Maintenance Training 55

4 Work Order Systems 65

5 Maintenance Planning and Scheduling Programs 76

6 Preventive Maintenance 98

7 Maintenance Inventory and Purchasing 114

8 Management Reporting and Analysis 129

9 World Class Maintenance Management 152

10 Integration of Maintenance Management 163

Index 169

Introduction—Status of Maintenance in the United States

In the United States, it was estimated that in 1979 there were over 200 billion dollars spent on maintenance. This is a sizable figure in anyone's estimation. However, more disturbing than the amount was the fact that approximately one-third of that total was spent unnecessarily. As we continue to the present, there have been no significant changes in maintenance policy, indicating the waste trend is probably still about one-third.

The largest change in the maintenance costs is the amount. Since 1979 maintenance costs have risen between 10 and 15% per year. If this amount is calculated, it is very likely that the maintenance expenditures in the United States are now over one-half a trillion dollars per year. If the waste ratio is holding steady, and we have no reason to believe that it is not, we could now be wasting as much per year as we were spending to do maintenance 10 years ago. This is illustrated in Fig. 1.

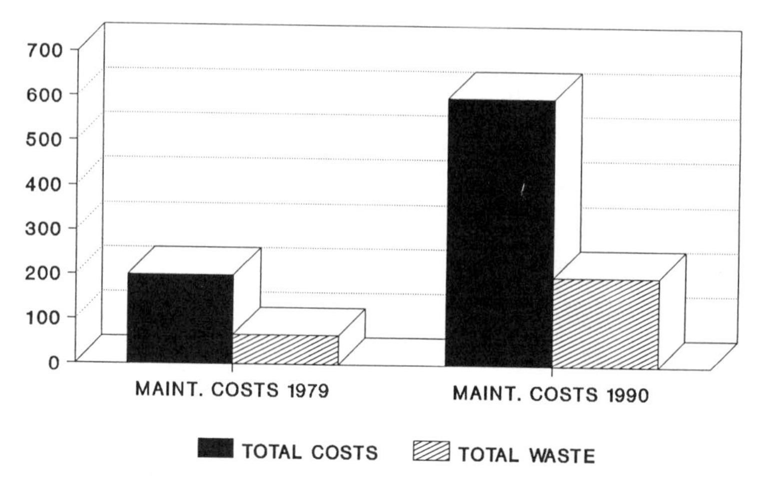

Figure 1. Maintenance costs comparison (projected).

Where do these wastes occur in maintenance? How can they be controlled? These questions can best be answered by looking at some statistics.

1. Less than 4 hr/day (out of a possible 8) are spent by maintenance craftsmen performing hands-on work activities (see Fig. 2). This fact is even more alarming when it is realized that the majority of maintenance organizations are performing as few as 2 hr of hands-on work. It is not that these individuals are lazy or shirking job responsibilities. It is the fact that they are not provided the necessary resources by management to perform the assigned job tasks. Providing these resources becomes important to increasing maintenance productivity and producing a substantial maintenance labor savings. If we would view maintenance salaries as a resource, we are paying (for example) $20.00 per hour and only utilizing this resource at a 50% level; that is a tremendous cost waste. We will examine in a later section methods that may be used to increase productivity.

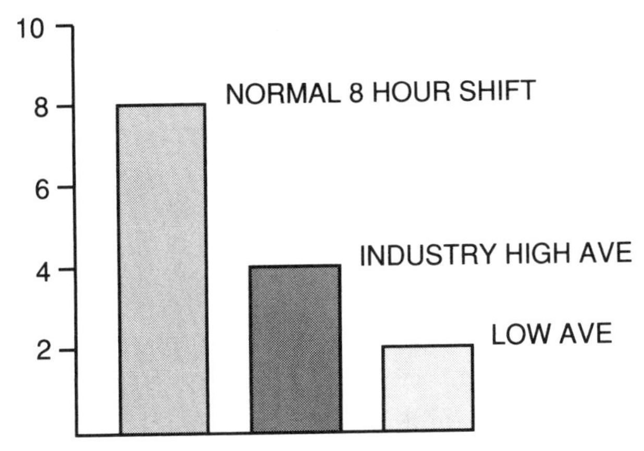

Figure 2. Actual hands-on work performed by maintenance craftsmen.

2. Only about one-third of all maintenance organizations use a job planner to plan maintenance activities (see Fig. 3). Most experts agree this is one of the largest potentials for cost savings in the maintenance arena. It is estimated that planned versus unplanned work may have a cost ratio as high as 1 : 5. Performing a $100.00 planned job could save as much as $400.00 over performing the same job in an unplanned mode. We will explore

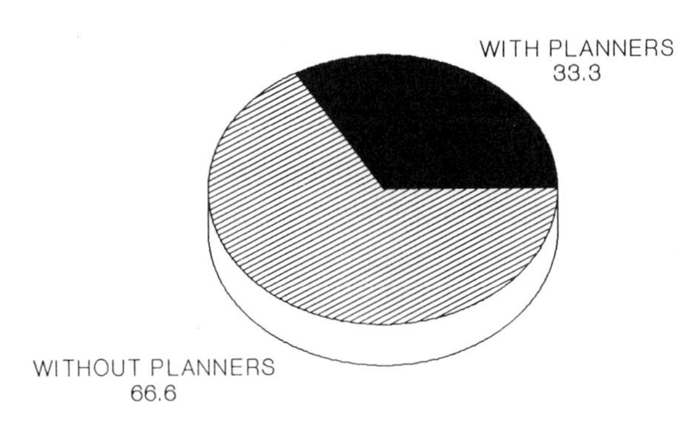

Figure 3. Maintenance planners—perspective.

maintenance planners and their qualifications and assignments in a later section.

3. The majority of all maintenance organizations either are dissatisfied with or do not have work order systems (see Fig. 4). This is one of the most critical indicators of the status of a maintenance organization. If a maintenance organization does not have a work order system in place that works properly, it is impossible to measure or control maintenance activities. The importance of a maintenance work order system and how to set up and use the system will be discussed in a following section.

4. Of companies that have work order systems (one-third of all companies) only about one-third track the work orders in a craft backlog format (actual 10% of total organizations) (see Figs. 5 and 6). This format will permit the manager to make logical staffing decisions based on how much work is projected for each craft. It will be a departure from the "I think I have enough craftworkers" or "let's work overtime to get caught up" modes that most organizations find themselves in today. Being able to justify employment levels to upper management is a necessary function of good maintenance management. We will discuss backlogs and their significance in a later section.

5. Of companies that have work order systems, only one-third compared their estimates of the work order labor and materials to the actual figures (see Fig. 7). Again, this means only about 10% of all organizations carry out some form of performance monitoring. Successful maintenance management requires performance monitoring. Proper methods of performance monitoring including maintenance analysis will be covered in a later section.

6. Of the companies with work order systems that allow for feedback, only one-third, again about 10% of all companies, perform any failure analysis on their breakdowns (see Fig. 8). Most

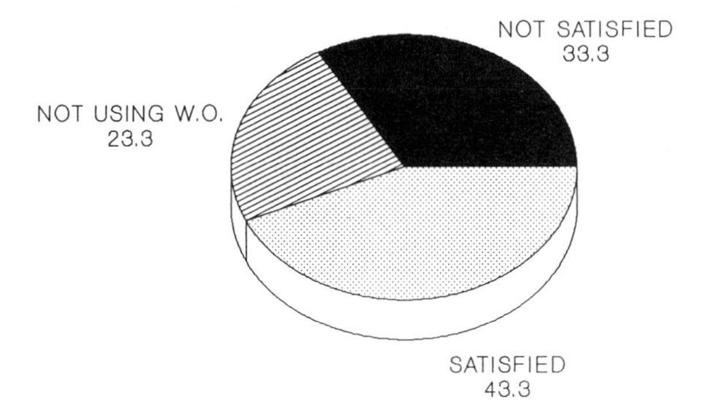

Figure 4. Analysis of the work order system.

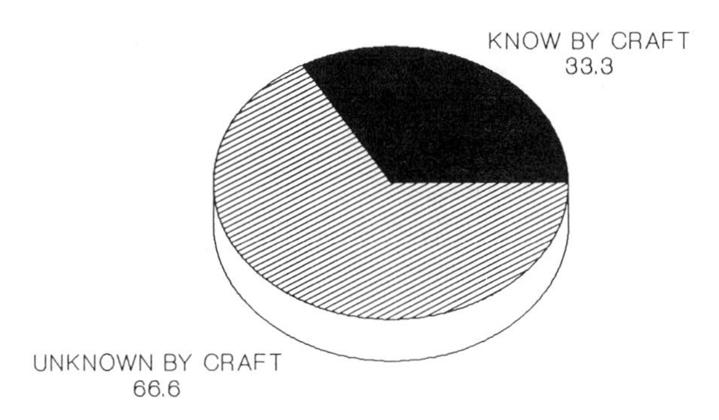

Figure 5. Craft backlogs (man-hr/craft).

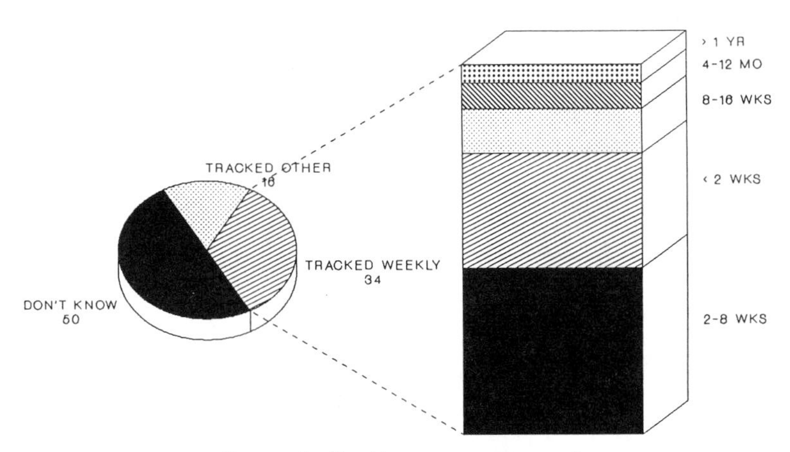

Figure 6. Backlog measurement.

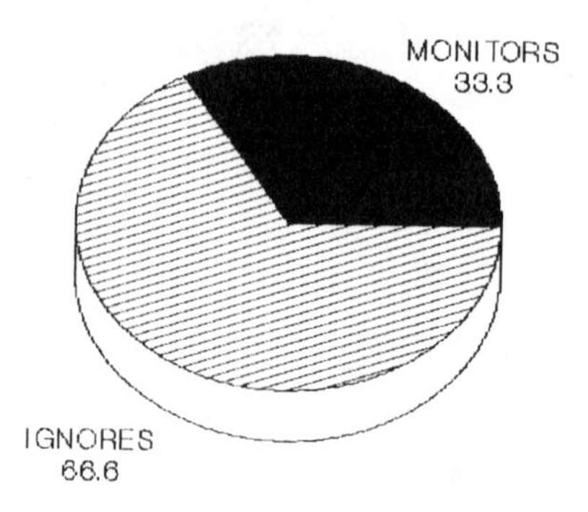

Figure 7. Performance monitoring.

of the other companies are just parts changers. For an operation to be cost effective, good practice in failure analysis must be followed. We will discuss this topic in a later section.

7. Overtime, another key indicator, in the United States averages about 14.1% of the total time worked by maintenance organizations (see Fig. 9). This figure is almost three times what it should be. Since maintenance is working so much overtime, it again indicates the reactive situation that is standard in the industry. Reducing overtime is essential if a maintenance organization is to be truly cost effective. We will examine proven methods for reducing overtime in later sections.

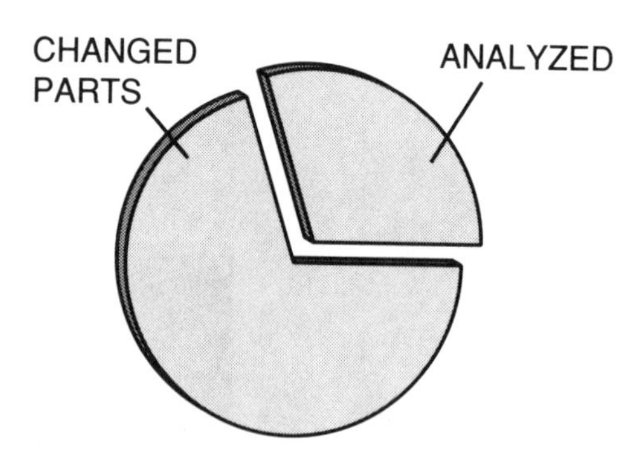

Figure 8. Failure analysis.

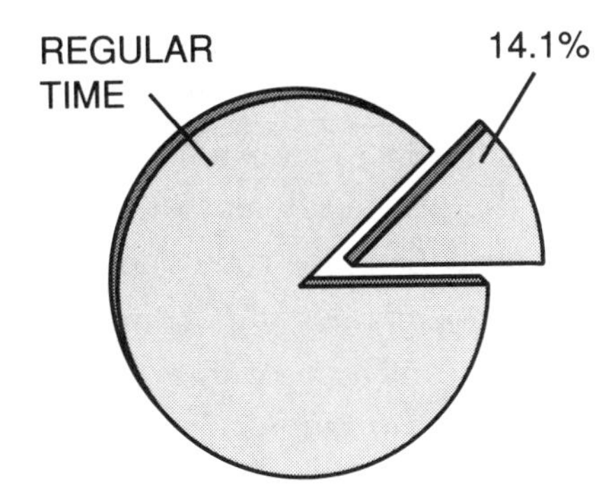

Figure 9. Overtime.

8. Preventive maintenance, another major part of any successful maintenance program, is presently satisfying the needs of about 22% of the maintenance organizations (see Fig. 10). This again illustrates major problems for the maintenance organizations. Without successful preventive maintenance programs, maintenance can only react to given situations. Preventive maintenance allows the organization to plan better and reduce maintenance costs. Over three-fourths of the organizations need major improvements in this area. We will discuss methods of implementing and improving preventive maintenance programs in a later section.

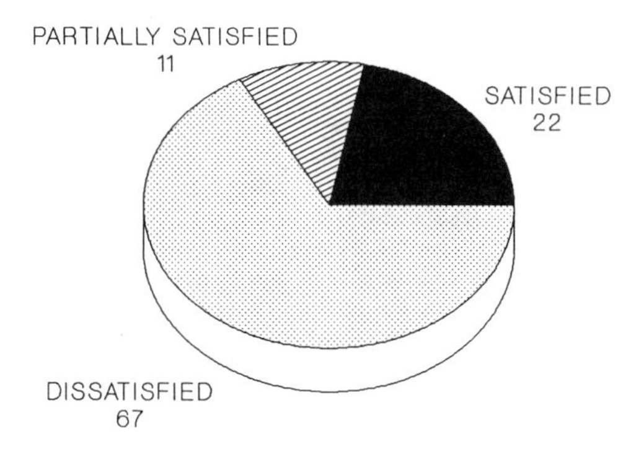

Figure 10. Preventive maintenance programs.

9. Related to preventive maintenance, almost three-fourths of the organizations have some form of lube routes and procedures (see Fig. 11). While this fact seems to be positive on the surface, it is not. Many of the organizations feel that preventive maintenance is nothing more than lube routes and procedures. So once they have these developed, they stop. However, preventive maintenance encompasses much more than lube routes. To be successful, maintenance organizations must go beyond the preliminaries and fully develop preventive maintenance programs.

10. One final fact related to preventive maintenance is the lack of coordination between operations/facilities and maintenance. Almost three-fourths of all organizations experience problems in coordinating preventive maintenance with the operations/facilities group (see Fig. 12). The problem is with communication. Either the maintenance organization has not communicated the need for the preventive maintenance or the operations/facilities group is not listening. Good, credible communication must be established if preventive maintenance is to be effective.

11. Second only to maintenance labor is the cost of maintenance materials. Depending on the type of operation/facility maintenance materials can range between 20 and 70% of the

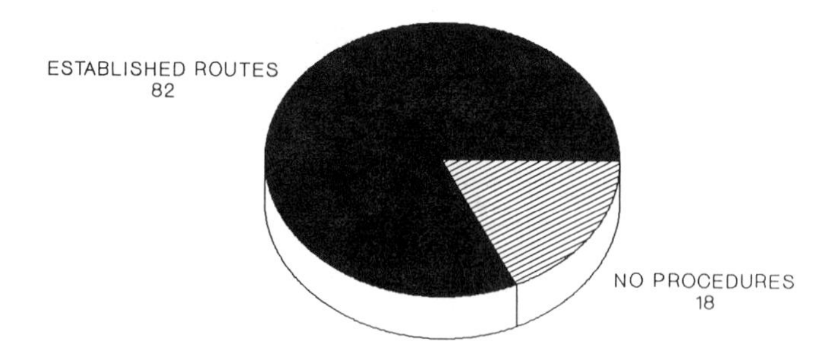

Figure 11. Lubrication routes.

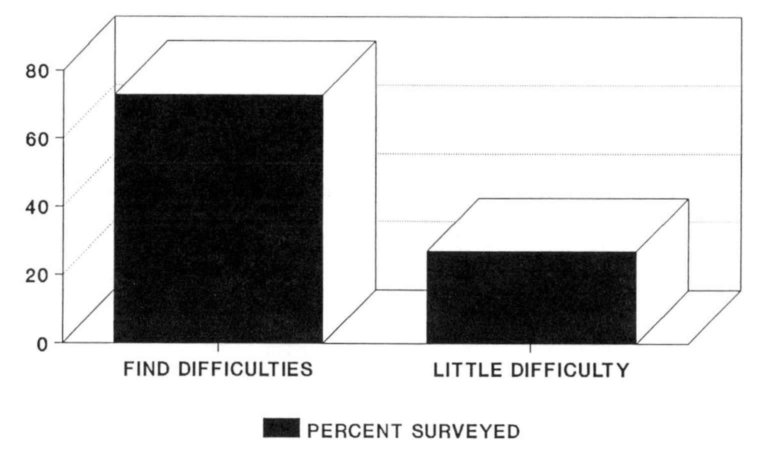

Figure 12. Coordination problems for preventive maintenance.

maintenance budget (see averages in Fig. 13). To manage maintenance successfully, materials must be given close scrutiny. We will examine in detail the needs maintenance has to properly manage their materials.

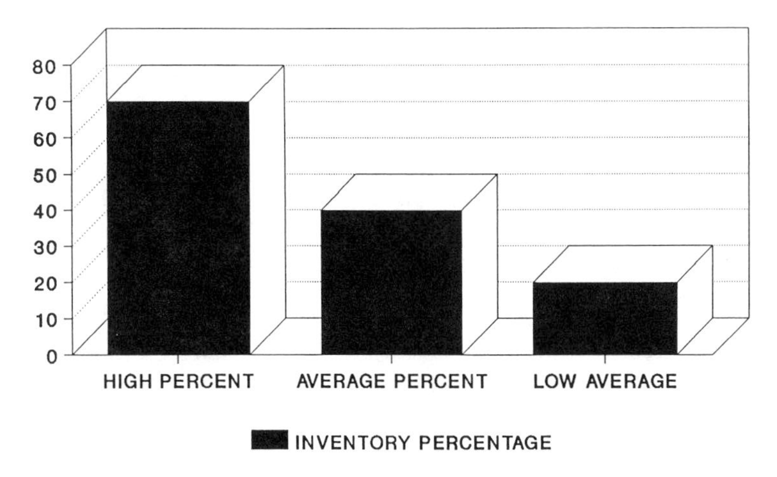

Figure 13. Maintenance inventory costs as a percentage of budget.

12. Many companies try to remedy maintenance materials problems by overstocking the storeroom. This is a problem because most will not take into account that inventory carrying costs are over 30% of the price of the items per year. So, for example, the costs of carrying a 1 million dollar inventory is over $300,000.00 per year. It is for this reason the inventories must be kept low, while still providing for a satisfactory level of service.

13. A third point of concern for maintenance materials is that maintenance is only responsible for their inventory in about 50% of the organizations. This means the other 50% of the time, someone else is telling maintenance what they should stock and how many they can stock or how many they can issue. It is the classic story of the tail wagging the dog.

14. While most will thus agree that maintenance costs are high, they do not know how high they are for their own site. In most cases the costs of maintenance repairs are calculated as the cost of maintenance labor and the maintenance materials to effect the repair. What larger figure that is not added is the cost of

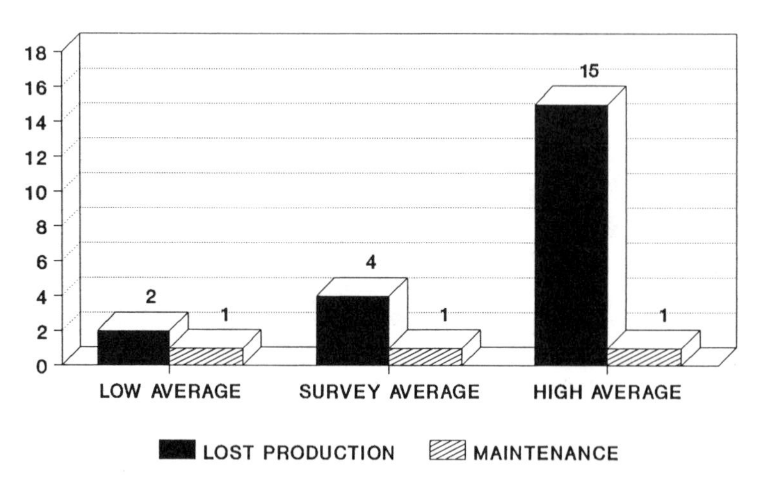

Figure 14. Lost production/maintenance cost comparision.

lost production. The range for this cost may be from 2 to 15 times the cost of the maintenance repair as illustrated in Fig. 14. The average is usually 4 to 1. So while a maintenance repair in labor and materials may be $10,000.00, the actual cost is really closer to $50,000.00. In fact, it is estimated that in the next several years the annual maintenance costs will exceed the amount spent on yearly new capital investment. If one realizes the scrutiny given to yearly capital expenditures, is it any wonder that maintenance expenditures are being closely examined.

1 Analyzing Maintenance Management

This chapter presents a method for analyzing the maintenance organization. The analysis is organized into sections corresponding to the remaining chapters in this book. For an explanation of the benchmarks given in the analysis, reference should be made to the appropriate chapters.

Part 1. Maintenance Organizations (Chapter 2)

1. Maintenance organizational chart:
   a. current and complete—4 pts
   b. not complete or over 1 year old—3 pts
   c. not current and not complete—2 pts
   d. none—0 pts
2. Job descriptions are available for:
   a. all maintenance positions (including supervisors)—4 pts

b. all maintenance positions (except supervisors)—3 pts
c. all maintenance supervisors (and no others)—2 pts
d. less than 50% of all maintenance positions—1 pt
e. no job descriptions—0 pts

3. Maintenance supervisor to hourly maintenance employee ratio:
   a. 8–12 to 1—4 pts
   b. 12–16 to 1—3 pts
   c. less than 8 to 1—2 pts
   d. more than 16 to 1—1 pt
   e. no shift maintenance supervisor—0 pts

4. Maintenance planner to hourly maintenance employee ratio:
   a. 15–20 to 1—4 pts
   b. 10–15 to 1—3 pts
   c. 20–25 to 1—2 pts
   d. 25–30 to 1—1 pt
   e. no planner or any other ratio than above—0 pts

5. Maintenance organizational assignments:
   a. responsibilities fully documented—4 pts
   b. responsiblities clear, good coverage, good dispatching—3 pts
   c. informal supervision and coordination, some gaps in job coverage—2 pts
   d. maintenance reports to production/operation—1 pt
   e. unclear lines of authority, jurisdictional—0 pts

6. Maintenance organization effort and attitude:
   a. excellent, pride in workmanship at all levels—4 pts
   b. steady work rate, professional operation—3 pts
   c. average work pace, only a few complaints—2 pts
   d. only occasional good efforts, frequent job delays, many complaints—1 pt

e. Constant disagreements within maintenance organization and between maintenance and operations /production/facilities—0 pts

7. Maintenance shop/work area locations:
   a. perfect—4 pts
   b. good (some improvement possible)—3 pts
   c. fair (major improvement possible)—2 pts
   d. poor (major improvement required)—1 pt
   e. unsuitable or nonexistent—0 pts

8. Maintenance shop/work area layouts:
   a. perfect—4 pts
   b. good (some improvement possible)—3 pts
   c. fair (major improvement possible)—2 pts
   d. poor (major improvement required)—1 pt
   e. unsuitable or nonexistent—0 pts

9. Maintenance tools/equipment quality and quantity:
   a. perfect—4 pts
   b. good (some improvement possible)—3 pts
   c. fair (major improvement possible)—2 pts
   d. poor (major improvement required)—1 pt
   e. unsuitable or nonexistent—0 pts

10. What percentage of maintenance personnel are tied to a pay incentive plan based on output?
    a. all—4 pts
    b. greater than 90%—3 pts
    c. greater than 75%—2 pts
    d. greater than 50%—1 pt
    e. less than 50% or other—0 pts

Part 2. Training Programs in Maintenance (Chapter 3)

1. Supervisory training:
   a. all are trained when salaried and additional training is mandatory on a scheduled basis—4 pts
   b. all are trained when salaried and additional training is offered on an optional basis—3 pts
   c. the majority are trained when salaried—2 pts
   d. the majority are offered and attend training offered on an infrequent or irregular basis—1 pt
   e. few are given initial training and little or no additional training is provided—0 pts

2. Planner training:
   a. all planners/schedulers have been to one or more public seminars providing instruction on maintenance planning and scheduling—4 pts
   b. all planners/schedulers are provided with a written training program for maintenance planning—3 pts
   c. all planners/schedulers receive one-on-one on the job training for at least 1 month—2 pts
   d. planner/scheduler training is on the job—1 pt
   e. there is no planner/scheduler training program—0

3. Details of planner training subjects (add 1 pt for each of the subjects covered; add 0 pts if there is no planner training program):
   a. work order planning and execution
   b. material planning
   c. scheduling practices
   d. project planning

4. General quality and productivity training:
   a. includes upper management, line supervision, hourly worker, support personnel—4 pts

b. includes upper management, line supervision, hourly workers—3 pts
c. includes upper management, line supervision—2 pts
d. is only for upper management—1 pt
e. no training program—0 pts

5. Maintenance craft training:
   a. training is tied to a pay and progression program—4 pts
   b. formal job experience is required before employment and on the job training is provided—3 pts
   c. formal job experience is required before hire—2 pts
   d. training is provided by on the job experience after hire—1 pt
   e. there is no formal training requirements for hire and no subsequent training is provided—0 pts

6. Maintenance training intervals: Formal maintenance training is provided to *all* maintenance craft employees at the frequency of:
   a. less than 1 year—4 pts
   b. between 12 and 18 months—3 pts
   c. between 18 and 24 months—2 pts
   d. not provided to all employees but to some in any of the above frequencies—1 pt
   e. no training is offered—0 pts

7. Format of maintenance training:
   a. training is a mix of classroom and lab excercises—4 pts
   b. training is all classroom—3 pts
   c. training is all in lab or workshop environment—2 pts
   d. training is all on the job—1 pt
   e. no formal craft training program exists—0 pts

8. Training program instructors:
   a. training is done by outside contract expert—4 pts
   b. training is done by staff subject expert—3 pts
   c. training is done by supervisors—2 pts
   d. training is done by hourly workers—1 pt
   e. training program does not exist—0 pts

9. The quality and skill level of the maintenance work force is:
   a. perfect—4 pts
   b. good (some improvement possible)—3 pts
   c. fair (major improvement possible)—2 pts
   d. poor (major improvement required)—1 pt
   e. unsuitable—0 pts

10. The quality and skill level of the supervisory group is:
   a. perfect—4 pts
   b. good (some improvement possible)—3 pts
   c. fair (major improvement possible)—2 pts
   d. poor (major improvement required)—1 pt
   e. unsuitable—0 pts

Part 3. Maintenance Work Orders (Chapter 4)

1. What percentage of maintenance man-hours are reported to a work order:
   a. 100%—4 pts
   b. 75%—3 pts
   c. 50%—2 pts
   d. 25%—1 pt
   e. less than 25%—0 pts

2. What percentage of maintenance materials are charged against a work order number when is?ed:
   a. 100%—4 pts

b. 75%—3 pts
c. 50%—2 pts
d. 25%—1 pt
e. less than 25%—0 pts

3. What percentage of total jobs performed by maintenance are covered by work orders:
   a. 100%—4 pts
   b. 75%—3 pts
   c. 50%—2 pts
   d. 25%—1 pt
   e. less than 25%—0 pts

4. What percentage of incomplete or backlog work orders are kept filed by equipment number:
   a. 100%—4 pts
   b. 75%—3 pts
   c. 50%—2 pts
   d. 25%—1 pt
   e. less than 25%—0 pts

5. What percentage of the work orders are filed by equipment number upon completion:
   a. 100%—4 pts
   b. 75%—3 pts
   c. 50%—2 pts
   d. 25%—1 pt
   e. less than 25%—0 pts

6. What percentage of the work orders are available for historical data analysis:
   a. 100%—4 pts
   b. 75%—3 pts
   c. 50%—2 pts
   d. 25%—1 pt
   e. less than 25%—0 pts

7. What percentage of the work orders are checked by the supervisor for work quality and completeness:
   a. 100%—4 pts
   b. 75%—3 pts
   c. 50%—2 pts
   d. 25%—1 pt
   e. less than 25%—0 pts

8. What percentage of the work orders are closed within 8 weeks from the date requested:
   a. 100%—4 pts
   b. 75%—3 pts
   c. 50%—2 pts
   d. 25%—1 pt
   e. less than 25%—0 pts

9. What percentage of the work orders are generated from the preventive maintenance inspections:
   a. 80–100%—4 pts
   b. 60–80%—3 pts
   c. 40–60%—2 pts
   d. 20–40%—1 pt
   e. less than 20%—0 pts

10. Add one point for each of the categories you track by work orders:
    a. required downtime
    b. required craft hours
    c. required materials
    d. requestor's name

Part 4. Maintenance Planning and Scheduling (Chapter 5)

1. What percentage of nonemergency work orders are completed within four weeks of the initial request?

a. more than 90%—4 pts
b. between 75 and 90%—3 pts
c. between 60 and 75%—2 pts
d. between 40 and 75%—1 pt
e. less than 40%—0 pts

2. Work order planning: add one point for each of the following:
   a. craft required
   b. materials required
   c. tools required
   d. specific job instructions or job plan

3. Percentage of planned work orders experiencing delays due to poor or incomplete plans:
   a. less than 10%—4 pts
   b. between 10 and 20%—3 pts
   c. between 20 and 40%—2 pts
   d. between 40 and 50%—1 pt
   e. more than 50%—0 pts

4. Who is responsible for planning the work orders:
   a. a dedicated maintenance planner—4 pts
   b. a maintenance supervisor—2 pts
   c. each craftworker—0 pts

5. Maintenance job schedules are issued:
   a. weekly—4 pts
   b. biweekly—3 pts
   c. between 3 and 7 days—2 pts
   d. daily—1 pt
   e. any other frequency—0 pts

6. The maintenance and production/facilities scheduling meeting is held:
   a. weekly—4 pts

b. biweekly—3 pts
c. between 3 and 7 days—2 pts
d. daily—1 pt
e. any other frequency—0 pts

7. The backlog of maintenance work is available by (add one point for each catagory):
   a. craft required
   b. department/area requesting
   c. requestor
   d. date needed by

8. When the job is completed, the actual time, material, downtime, and other information is reported by:
   a. the craftsmen performing the work—4 pts
   b. the supervisor of the group—3 pts
   c. anyone else—2 pts
   d. information is not recorded—0 pts

9. What percentage of the time are the actuals compared to the estimates for monitoring planning effectiveness:
   a. more than 90%—4 pts
   b. between 75 and 90%—3 pts
   c. between 60 and 75%—2 pts
   d. between 40 and 75%—1 pt
   e. less than 40%—0 pts

10. What is the reporting relationship between planners and supervisors:
   a. both report to the same maintenance manager—4 pts
   b. the planner reports to the supervisor—3 pts
   c. the supervisor reports to the planner—2 pts
   d. the supervisor and planner report to operations/facilities—0 pts

Part 5. Preventive Maintenance (Chapter 6)

1. The preventive maintenance (PM) program includes (add 1 point for each type included):
   a. lubrication checklists
   b. detailed inspection checklists
   c. personnel specifically assigned to the PM program
   d. PM diagnostics such as vibration analysis, oil sample analysis, infrared heat monitors, etc.

2. What percentage of the PM inspection/task checklists are checked to ensure completeness:
   a. more than 90%—4 pts
   b. between 75 and 90%—3 pts
   c. between 60 and 75%—2 pts
   d. between 40 and 75%—1 pt
   e. less than 40%—0 pts

3. What percentage of the plant critical equipment is covered by a PM program:
   a. more than 90%—4 pts
   b. between 75 and 90%—3 pts
   c. between 60 and 75%—2 pts
   d. between 40 and 75%—1 pt
   e. less than 40%—0 pts

4. What percentage of the PM program is checked against an equipment item's history annually to ensure good coverage?
   a. more than 90%—4 pts
   b. between 75 and 90%—3 pts
   c. between 60 and 75%—2 pts
   d. between 40 and 75%—1 pt
   e. less than 40%—0 pts

5. What percentage of the PM's are completed within 1 week of the due date?
   a. more than 90%—4 pts
   b. between 75 and 90%—3 pts
   c. between 60 and 75%—2 pts
   d. between 40 and 75%—1 pt
   e. less than 40%—0 pts

6. What determines the frequency of a PM inspection or task/service interval?
   a. program is condition-based—4 pts
   b. program is based on a combination of equipment run time and fixed calendar interval—3 pts
   c. program is based on equipment run time only—2 pts
   d. program is based on calendar intervals—1 pt
   e. program is dynamic and is scheduled based on completion date of previous task—0 pts

7. What percentage of the inspections/tasks have more than five lines of detail or instruction?
   a. more than 90%—4 pts
   b. between 75 and 90%—3 pts
   c. between 60 and 75%—2 pts
   d. between 40 and 75%—1 pt
   e. less than 40%—0 pts

8. The average time to complete a PM inspection or task is:
   a. 4 hr—4 pts
   b. 4–8 hr—3 pts
   c. 2–4 hr—2 pts
   d. less than 2 hr—1 pt
   e. any other time—0 pts

9. PM actuals and results are checked annually for time and material estimate accuracy on what percentage of the program?

a. more than 90%—4 pts
b. between 75 and 90%—3 pts
c. between 60 and 75%—2 pts
d. between 40 and 75%—1 pt
e. less than 40%—0 pts

10. Who is responsible for performing PM tasks?
a. dedicated PM personnel—4 pts
b. specific individuals on each crew—3 pts
c. any individual on a crew—2 pts
d. entry level craftworkers—1 pt
e. operating personnel—0 pts

Part 6. Maintenance Inventory and Purchasing (Chapter 7)

1. What percentage of the time are materials in stores when required by the maintenance organization?
a. more than 95%—4 pts
b. between 80 and 95%—3 pts
c. between 70 and 80%—2 pts
d. between 50 and 70%—1 pt
e. less than 50%—0 pts

2. What percentage of the items in inventory appear in the maintenance stores catalog?
a. more than 90%—4 pts
b. between 75 and 90%—3 pts
c. between 60 and 75%—2 pts
d. between 40 and 75%—1 pt
e. less than 40%—0 pts

3. Who controls what is stocked as maintenance inventory items?
a. maintenance—4 pts
b. anyone else—0 pts

4. The maintenance stores catalog is produced in:
   a. alphabetic and numeric listings—4 pts
   b. alphabetic only—2 pts
   c. numeric only—2 pts
   d. catalog is incomplete or nonexistent—0 pts

5. The aisle/bin location is specified for what percentage of the stores items?
   a. more than 95%—4 pts
   b. between 90 and 95%—3 pts
   c. between 80 and 90%—2 pts
   d. between 70 and 80%—1 pt
   e. less than 70%—0 pts

6. What percentage of the maintenance stores items are issued to a work order or account number upon leaving the store?
   a. more than 95%—4 pts
   b. between 90 and 95%—3 pts
   c. between 80 and 90%—2 pts
   d. between 70 and 80%—1 pt
   e. less than 70%—0 pts

7. Maximum and minimum levels for the maintenance stores items are specified for what percentage of the inventory?
   a. more than 95%—4 pts
   b. between 90 and 95%—3 pts
   c. between 80 and 90%—2 pts
   d. between 70 and 80%—1 pt
   e. less than 70%—0 pts

8. A reorder list is sent to purchasing:
   a. daily—4 pts
   b. every 1–3 days—3 pts
   c. weekly—2 pts
   d. any other frequency—0 pts

9. Maintenance stores inventory levels are updated daily upon receipt of materials what percentage of the time?
   a. more than 95%—4 pts
   b. between 90 and 95%—3 pts
   c. between 80 and 90%—2 pts
   d. between 70 and 80%—1 pt
   e. less than 70%—0 pts

10. What percentage of the items are checked for at least one issue every 6 months?
    a. more than 90%—4 pts
    b. between 80 and 90%—3 pts
    c. between 70 and 80%—2 pts
    d. between 50 and 70%—1 pt
    e. less than 50%—0 pts

Part 7. Maintenance Reporting (Chapter 8)

1. What percentage of the time are the maintenance reports distributed on a timely basis to the appropriate personnel?
   a. more than 90%—4 pts
   b. between 75 and 90%—3 pts
   c. between 60 and 75%—2 pts
   d. between 40 and 75%—1 pt
   e. less than 40%—0 pts

2. What percentage of the time are the reports distributed within 1 day of the end of the time period specified in the report?
   a. more than 90%—4 pts
   b. between 75 and 90%—3 pts
   c. between 60 and 75%—2 pts
   d. between 40 and 75%—1 pt
   e. less than 40%—0 pts

3. Add one point for each of the following equipment reports you produce:
   a. equipment downtime in order of highest to lowest total hours (weekly or monthly)
   b. equipment downtime in order of highest to lowest in total lost production dollars (weekly or monthly)
   c. maintenance cost for equipment in order of highest to lowest (weekly or monthly)
   d. MTTR and MTBF for equipment

4. Add one point for each of the following PM reports you produce:
   a. PM overdue report in order of oldest to most recent
   b. PM cost per equipment item in descending order
   c. PM hours versus total maintenance hours per item, expressed as a percentage
   d. PM costs versus total maintenance costs per equipment item, expressed as a percentage

5. Add one point for each of the personnel reports you produce:
   a. time report showing hours worked by employee divided by work order
   b. time report showing hours worked by craft in each department/area
   c. time report showing total hours spent by craft on emergency/preventive/normal work
   d. time report showing total overtime hours compared to regular hours

6. Add one point for each of the planning reports you produce:
   a. total work order costs estimated versus total work order actual costs by individual work order, by supervisor, or by craft

b. a backlog report showing the total hours ready to schedule versus the craft capacity per week
c. a planning efficiency report showing the hours and materials planned versus the actual hours and materials used per work order
d. a planning effectiveness report showing the number of jobs closed out that were 20% over or under the labor or material estimates by planner and supervisor

7. Add one point for each of the scheduling reports you produce:
   a. hours worked as scheduled compared to actual hours worked
   b. weekly crew or craft capacity averaged for the last 20 weeks
   c. total number of maintenance work orders scheduled compared to the actual number of work orders completed
   d. number of work orders spent on preventive maintenance compared to emergency maintenance and normal maintenance

8. Add one point for each of the inventory reports you produce:
   a. stock catalog by alphabetical and numerical listing
   b. inventory valuation report
   c. inventory performance report showing stockouts and level of service, turnover rate, etc.
   d. inventory where used report

9. Add one point for each of the purchasing reports you produce:
   a. vendor performance showing promised and actual delivery dates
   b. price performance, showing the quoted and actual prices

c. buyer or purchasing agent performance report
d. nonstock report showing all direct buys for items not carried in stock for a specified period

10. Add one point for each administrative report you produce:
    a. monthly maintenance costs versus monthly maintenance budget with a year to date total
    b. comparison of labor and material costs as a percentage of total maintenance costs
    c. total costs of outside contractor usage broken down by contractor/project
    d. maintenance cost per unit of production (or by square foot for facilities)

Part 8. Maintenance Automation (Chapter 9)

1. What percentage of all maintenance operations are computerized?
    a. more than 90%—4 pts
    b. between 75 and 90%—3 pts
    c. between 60 and 75%—2 pts
    d. between 40 and 75%—1 pt
    e. less than 40%—0 pts

2. What percentage of maintenance activities are planned and scheduled through the system?
    a. more than 90%—4 pts
    b. between 75 and 90%—3 pts
    c. between 60 and 75%—2 pts
    d. between 40 and 75%—1 pt
    e. less than 40%—0 pts

3. What percentage of the maintenance inventory and purchasing functions are performed in the system?

a. more than 90%—4 pts
b. between 75 and 90%—3 pts
c. between 60 and 75%—2 pts
d. between 40 and 75%—1 pt
e. less than 40%—0 pts

4. Is there an existing interface between the maintenance information system and the production scheduling system?
   a. yes—4 pts
   b. no—0 pts

5. Is the information in the integrated system reliable and accurate?
   a. yes—4 pts
   b. no—0 pts

6. Is there an interface between the payroll system and the timekeeping system in the maintenance information system?
   a. yes—4 pts
   b. no—0 pts

7. Is there an interface between the maintenance system and the accounting system?
   a. yes—4 pts
   b. no—0 pts

8. What percentage of the maintenance personnel are using the system for their job functions with a high level of proficiency?
   a. more than 90%—4 pts
   b. between 75 and 90%—3 pts
   c. between 60 and 75%—2 pts
   d. between 40 and 75%—1 pt
   e. less than 40%—0 pts

9. Is the maintenance organization consulted when any corporate policy affecting them is made?
   a. yes—4 pts
   b. sometimes—2 pts
   c. no—0 pts

10. Does a cooperative spirit exist at all levels of the corporate structure allowing maintenance to contribute to the overall profitability of the organization?
   a. yes—4 pts
   b. in most cases—3 pts
   c. in few cases—1 pt
   d. no—0 pts

Part 9. The Analysis

Total points possible	320
Your score	___
320 — 288 pts	You qualify for world class
288 — 256 pts	Close, examine areas where you lost points
256 — 224 pts	You need adjustments in several areas, examine your sectional totals to find weak spots and prioritize
Less than 224	Examine your goals and priorities and consider the points in this book's chapters where you want to improve

2 Maintenance Organizations

Properly determining the type of maintenance organization is influenced by the goals and objectives of the maintenance organization. If the goals and objectives are progressive and the maintenance organization is recognized as a contributor to the corporate bottomline, variations on some of the more conventional organizational structures can be used. The typical goals and objectives for a maintenance organization (listed in Fig. 2-1) are as follows:

1. *Maximum production or availability of facilities at the lowest cost and at the highest quality and safety standards.* This first statement seems to be all encompassing. However, it can be divided into some smaller components:

a. *Maintaining existing equipment and facilities.*

This is the primary reason for the existence of the maintenance organization. Unless the equipment or facilities owned by the corporation are operating or functional, there is no advantage

* Maximum production at the lowest cost, the highest quality, and the optimum safety standards.
* Identify and implement cost reductions.
* Provide accurate equipment maintenance records.
* Collect the necessary maintenance cost information.
* Optimize maintenance resources.
* Optimize capital equipment life.
* Minimize energy usage.
* Minimize inventory on-hand.

Figure 2-1. Objectives of maintenance management.

in having them. The first subobjective is the keep-it-running charter of maintenance.

b. *Equipment and facilities inspections and services.*

This is generally referred to as the preventive/predictive maintenance program. This activity is designed to increase the availability of the equipment/facilities by reducing the number of unexpected breakdown or service interruptions. It is one of the most critical parts of any program designed to achieve the first main goal of "keeping-it-running."

c. *Equipment installations or alterations.*

This is not the responsibility of maintenance in all organizations, since installations or alterations are usually performed by outside contract personnel. However, maintenance must still maintain the equipment, so they should be involved in any alterations or new equipment installations.

In viewing these three subgoals, the maintenance organizations will always attempt to maximize the company's resources, keeping the overall costs as low as possible, while ensuring the safety of personnel and the quality of the product/facilities.

2. *Identify and implement cost reductions.* This objective is for the maintenance organizations to find ways to decrease maintenance and operations expenses. There are many ways this may be accomplished. For example, by examining maintenance practices, it is common to find that adjustments made in

tools

training

repair procedures

work planning

can all reduce the amount of labor or materials that may be required to perform a specific job. In addition, any time gained while making repairs translates into reduced downtime (or increased availability), which is more costly than maintenance expenditures.

When there are adjustments made to reduce costs, studies need to be conducted to show the before and after results. The quantifying of the improvements is critical to maintaining management support for maintenance activities.

3. *Provide accurate equipment maintenance records.* This objective seems almost impossible at times because maintenance records are generally collected as work orders and then must be compiled into reports showing meaningful information or trends. The problem is finding enough time to put valuable information on each individual work order. Since most work is done in a reactionary mode, it is difficult to record events after the fact. For example, recording how many times a circuit breaker for a drive motor was reset in one week might seem somewhat insignificant to record on a work order. But, if the overload was due

to an increased load on the motor by a bearing wearing inside the drive, it could be checked and discovered before the equipment experienced a catastrophic failure. Accurate record keeping is important if maintenance is really going to maintain equipment.

4. *Collect necessary maintenance cost information.* This objective is also related to objective 3; however, cost information is more detailed than repair information. For example, cost information is divided into these general areas:

Labor costs

Material costs

Tool and equipment costs

Contractor costs

Lost production costs

Miscellaneous costs

The importance of collecting this information is highlighted in maintenance budgeting. If accurate cost histories are not collected, how can the manager budget what next year's expenses will be, or should be? It is difficult to express to plant management, "We want to reduce maintenance manpower by 10% for next year," when you do not really know how the manpower resource was allocated for this year. Also, if the manpower figures are only in dollar amounts, the differences in pay scales may make it difficult to determine how much manpower was used in total hours by craft. The information must be collected in dollars and in hours by craft. Where is the information collected?

Collecting the cost information is again tied to work order control. Knowing both the hours spent on the work order and the labor rates of the individuals performing the work allows a calculation of the labor used for the work order. Adding up these charges over a given time period for all work orders gives you the total labor used. If you were to add up the hours spent by each craft, you could also learn about the manpower resources necessary.

Material costs can also be determined by tracking what parts were used on the job to each work order. The number of parts times their dollar value (obtained from stores or purchasing) allows the calculation of the total material dollars spent for a given time period.

Contractor and other costs information must also be collected at a work order level. Each work order form should have the necessary blanks for filling in this information. Only by tracking information at the work order level can you roll up costs from equipment, to line, to department, to area, to total plant. Collecting the information at this level also allows for costs for equipment types, maintenance crafts, cost centers, and so on. By utilizing the data gathered through the work order, detailed maintenance analysis can be performed, as will be discussed in a subsequent chapter.

5. *Optimize maintenance resources.* Making the most with the resources at hand is important in maintenance. There are very few maintenance organizations that have as many people, materials, or tools as they could use. Since there is a shortage of maintenance resources, they must be used carefully. For example, it is estimated that by good planning and scheduling practices, a reactive maintenance organization may double the productivity of its employees. While this may seem to be a bold statement, it has been accomplished many times.

Maintenance supplies is another area of savings. When good controls are put into place, organizations have seen as much as a 15% reduction in inventory storage costs, while increasing the service level of the stores. These types of reductions, while improving service, are essential to optimizing the present resources.

Optimizing maintenance resources as highlighted in this section can only be achieved by good planning and scheduling practices. The disciplines necessary to develop these controls will be discussed in a subsequent chapter.

6. *Optimize capital equipment life.* Any equipment, whether a complex industrial robot or your automobile, requires maintenance if it is to deliver its desired service life. Industrial equipment has its life cycle optimized by an effective preventive maintenance program. Preventive maintenance (PM) is a neglected discipline in many organizations throughout North America. The interest shown by management in short-term gains has contributed to this neglect. Nevertheless, if capital equipment life is to be optimized, only effective preventive maintenance programs will achieve the goal. How to achieve this goal will also be covered in a subsequent chapter.

7. *Minimize energy usage.* While this objective may seem to be more production or operations oriented, it is maintenance related. How? Equipment and facilities that are properly maintained will require less energy to operate. For example, equipment with a poor maintenance schedule will have bearings without proper lubrication or adjustment, couplings not properly aligned, or gears misaligned, all of which contribute to poor performance and require more energy to operate. The key to achieving this objective is also a good preventive/predictive maintenance schedule. As was mentioned, how to set up an effective preventive maintenance program will be covered in a subsequent chapter.

8. *Minimize inventory on hand.* Since maintenance spare parts average 40% of the total maintenance budget and companies can actually have millions of dollars in inventory, reducing the on-hand quantities is a key objective. The costs of holding an inventory item in stock will average between 20% and 30% of the actual price of the item. So for companies with $2,000,000.00

in inventory, it is costing them between $400,000.00 to $600,000.00 per year to keep that inventory level. Obviously, any reduction in inventory results in compounded savings. When improving maintenance controls, this is the area showing quickest payback.

Management and Maintenance

During the last 20 years, management has become more production oriented and at times has sacrificed long-term benefits for short-term profitability. Foreign competitors have taken advantage of this trend to develop strategic plans, building strong, complete organizations. One of the foremost areas of development for them is the maintenance function. Maintenance is extremely important to being competitive in the world market.

But have U.S. companies followed their lead? The answer for most companies, sadly, is no. I personally have seen plants where the maintenance force one day is required to work on solid-state controls and the next day to perform janitorial service in the lavatories. It is difficult in this environment for maintenance personnel to develop a positive attitude of their value to the corporation.

If maintenance is to become a contributing factor to the survival of most U.S. corporations, management must change their view toward maintenance. If they do, they will achieve "world class" competitiveness. Three of the goals that will be achieve by the management are listed in Fig. 2-2.

Achieving the goals necessary to have a contributing maintenance organization will require some decisions to be made concerning the maintenance organization and the type of maintenance service to be provided in the operation. The types of decisions will be discussed, beginning with the type of service required from the equipment.

Management must understand maintenance management if

* They want to improve uptime, quality, and utilization
* They want to reduce maintenance repair, labor, and material costs
* They want to prepare the company organization to compete in the future

Figure 2-2

Equipment Service Level

Equipment service level indicates the amount of time the equipment is available for its intended service. The amount of service required from the equipment and its resultant costs determines the type of maintenance philosophy a company will adopt. The six typical types of philosophies are listed in Fig. 2-3.

1. **Breakdown maintenance** is just what its name implies—the equipment is run until it breaks down. There is no preventive maintenance; the technicians work only on equipment that is malfunctioning. This is the most expensive way to do maintenance. Equipment service level

* 100% breakdown
* Minor lube program
* Preventive maintenance
* Predictive maintenance
* Condition-based maintenance
* Zero failure maintenance

Figure 2-3. Types of maintenance.

is generally below acceptable levels with product quality usually impacted.

2. **Minor lube programs** are one step removed from breakdown maintenance programs. The equipment still is not overhauled until it breaks down, except that with the lube program it takes longer for the equipment to break down. Unfortunately, many companies mistake the lube program for a preventive maintenance program. Equipment service level is still not satisfactory under this program.

3. **Preventive maintenance** includes the lubrication program (from item 2) plus routine inspections and adjustments. This program allows many potential problems to be corrected before they occur. The methods for organizing and developing preventive maintenance programs will be detailed in a subsequent chapter. With this maintenance, equipment service levels begin to enter the acceptable range for most operations.

4. **Predictive maintenance,** another type of preventive maintenance, allows the forecasting of failures through analysis of the condition of the equipment. The analysis is generally conducted through some form of trending of a parameter such as vibration, temperature, or flow. Predictive maintenance allows equipment to be repaired at times that do not interfere with production schedules. This removes one of the largest factors from the downtime cost. The equipment service level will be very high under this type of maintenance.

5. **Condition-based maintenance** is maintenance performed as it is needed, with the equipment monitored continually. Some plants will have the PLCs (programmable logic controllers) wired directly to a computer to mon-

itor the equipment condition in a real time mode. Any deviation from the standard normal range of tolerances will cause an alarm (or, in some situations, a repair order) to be generated automatically. This real time trending allows for the maintenance to be performed in the most cost effective manner. This is the optimum maintenance cost versus equipment service level method available. However, the start-up and installation costs can be very high. Many companies are moving toward this type of maintenance.

6. **Zero failure maintenance** is used in any environment where the cost of a failure and the resulting production outage is very high. This type of maintenance combines several of the prior techniques to produce a maintenance environment where all critical points on equipment and processes are monitored in a real time mode with the data being charted and trended and projections made as to the service life remaining in each item. When the equipment or process becomes questionable, it is taken off-line and repairs are made; the equipment or process is then returned to service. While this type of maintenance is sophisticated, it is also the most expensive. It is only used in processes where the cost can be justified.

Maintenance Staffing Options

This is an area that deserves attention in a maintenance organization. There are four methods that are commonly used to staff the maintenance organization. These are listed in Fig. 2-4.

1. **Complete in-house staff** is the traditional approach in most U.S. companies. This is where the craftworkers performing the maintenance are direct employees of the company. The administrative functions for each employee are

* Complete in-house staff
* Combined in-house/contract staff
* Contract maintenance staff
* Complete contracting maintenance staff

Figure 2-4. Maintenance staffing options.

the responsibility of the company. The salary, benefits, etc., are also the responsibility of the company.

2. **Combined in-house/contract staff,** where the in-house staff will perform most of the maintenance, but contractors will perform certain maintenance tasks, has become a more common approach to maintenance in the 1980s. Examples of contract staff maintenance are service on air conditioners, equipment rebuilds, or insulation. This method can reduce the amount of staff required for specific skill functions. If the contract personnel are not required full time, this can contribute an added savings.

3. **Contract maintenance staffs** have company supervisors but use contract employees. This method is common in Japan and is gaining popularity in the United States. The contractor is responsible for providing the proper skilled individuals, which removes the burden of training and personnel administration from the company. One disadvantage is not having the same employees all of the time. In this situation you do lose some familiarity with the equipment, but the interaction between the in-house supervision and the contract personnel can help to compensate for this unfamiliarity.

4. **Complete contracting maintenance staff** includes all craftsmen, planners, and supervisors. The supervision generally will report to a plant engineer or plant manager.

This eliminates the need for any in-house maintenance personnel. While this program is not yet popular in the United States, when coupled with an operator-based preventive maintenance program (explained in Chapter 6), it can prove to be a cost effective and valid alternative to conventional maintenance organizations.

Geographical Maintenance Organizations

In addition to the types of maintenance organizations, there are also geographical divisions for the organization, regardless of how it is staffed. These divisions are listed in Fig. 2-5.

Centralized maintenance organizations have all members of the organization reporting to a central location for assignment. All work requests are turned into the central area for scheduling and dispatch. The utilization of the labor force is high, since there is always work for them to do because they are not assigned to a fixed geographical area. However, the drawback to this arrangement is in plants with a large area, response time to trouble calls can be long, contributing to increased equipment downtime.

Area maintenance organizations have small maintenance shops spread throughout the plant. A certain number of employees is assigned to each area, with supervisors assigned to cover one or more areas, depending on the number of employees. This arrangement remedies the slow response time for breakdowns, since the employees are physically close to the equipment. The

Centralized	Area	Combination
High utilization	Low utilization	Good utilization
Slow response	Quick response	Optimum response
Labor pool	Equipment ownership	

Figure 2-5. Maintenance organizations.

maintenance employees also develop some equipment "ownership," since they are assigned to the equipment. The problem with this arrangement is the labor utilization. If the craftworkers have the equipment running and are caught up on the preventive maintenance program, then they may have no work in their area. However, in another area, the craftworkers may be overloaded, with perhaps several equipment items broken down.

The dilemma faced by supervision is whether to pull craftworkers from other areas to cover the breakdowns, hoping the equipment will not breakdown in the area the craftworkers are being drawn from. The response time will be poor, the utilization is not good. If all areas have the same workload, the area concept can work. It takes the effort of a dedicated staff, willing to be flexible if optimum utilization is to be achieved.

Combination organizations are a compromise between the area and the centralized organizations. They are basically hybrids, incorporating the best of both organizations. The concept is to station several small groups of employees near critical equipment, while keeping the main group in a centralized area. This allows for the emergency activities to be handled quickly, while most of the employees can be used for larger or scheduled repair jobs.

Combination organizations appear to be the way of the future. In fact, some of the Japanese styles of maintenance fit very well into the combination organizations. For example, the concept of operator-based maintenance (TPM), in which the operators are responsible for the routine maintenance on the equipment they are assigned to operate. If any large problems arise, the maintenance craftworkers are called. This arrangement is very similar to the combination organization, with the operators taking the place of the area maintenance craftworkers.

Many of the points covered in this section have not included one of the most common considerations: union involvement. The strength and posture of the unions will help determine what management will be able to accomplish in changing or positioning the organization. Generally, a concise, clear presentation of the facts surrounding the change, including the benefits, will help the union see the need for any changes. Communication is vital if changes are to be successful.

Growing Maintenance Organizations

In examining maintenance organizations, we find they all have the same growth pattern. This pattern is pictured in Fig. 2-6.

When companies are small, they may only have one machine.

1. One production worker ... one machine. He runs and maintains ... big jobs sent outside
2. Several machines ... several productions workers ... One multiskilled maintenance worker
3. More production increase ... add maintenance people, also maintenance supervisor
4. Maintenance people will begin to specialize in types of repair
5. Craft lines develop
6. Central organization gets too large to manage, area organization develops
7. Development of central crafts and shops to support area crafts: like outside contractor
8. Area organizations tend to go back to multicraft concept
9. Labor pools are used to fill in when more manpower is needed

Figure 2-6. Organizational growth.

The operator of the machine will run and maintain it. Any small repairs or services are performed by the operator. If a large breakdown occurs, the machine is disassembled by the operator and the defective parts are sent out for repair.

As the company continues to grow, several machines are now added. This also necessitates the addition of several production workers. Since the production workers are not dependent on one machine any longer, the first maintenance worker is added. This individual will be multiskilled to be able to care for the variety of repairs that will be necessary.

The third step in the growth pattern will be the addition of more machines and production workers. This leads to the addition of more maintenance workers. With this level of manpower required by maintenance, it is no longer convenient to have the maintenance workers report to the production supervisor. At this level it is necessary to have a maintenance supervisor in place.

The fourth step in the growth pattern is to watch the maintenance personnel begin to specialize in their particular skill area. It begins by having craftworkers become proficient at repairing a particular piece of equipment or performing a particular type of repair. As the number of craftworkers continues to increase, the specialization continues.

The fifth step in the growth pattern is the development of craft lines. This may be due to union influence or the natural progression from step four. The lines can either be strict or informal, but will increasingly become distinct.

The sixth step in the growth pattern is the organization that becomes too large to be managed from a central location. There may be several factors that contribute to the management problem; the internal geography of the plant can be one factor. For example, if the plant covers several hundred acres, it may be physically impossible to manage it from one location, even with the help of radios, bicycles, or manned carts. This situation is when the organization is divided into the area concept and allows small maintenance departments, paralleling organizations in steps 1–3, to be utilized. This step leads to interesting internal growth.

Most organizations will develop two alternatives at this stage: further internal growth or outside contracting. Internal growth will develop central crafts or shops to support the area organizations. Thus, as in step 1, when the repair is too large or complicated, or it requires special equipment, it is sent to the central shops. As more work is required from the central shops, they tend to grow, while the area organizations tend to add employees only when new demands are made on their area (such as new equipment additions) or when attrition occurs.

Outside contracting occurs when the company either does not have the resources to implement central crafts or decides it is more cost effective to contract with a local shop for machining, rewinding, or installation. The determining factors here are skill level of the contractor's workforce, response time, and synergism between the contractor and the company.

In the final growth step, the eighth step, the area organizations will tend to return to the multicraft concept, allowing for the maximum flexibility of the labor resources assigned to an area. To assist in the peak work periods, it is possible for the central organization to maintain a pool of qualified individuals, the ninth step, proficient in various areas of the plant. There are companies with as many as 30 or 40 area organizations within a single plant, coupled with central organizations and outside contracting, managed by area organizations, reporting to a central maintenance management organization, that provide an optimum service/cost factor arrangement. Each company will make the policy decision several times before they find their optimum organization.

Maintenance Staffing

If a company has decided on its growth pattern and organizational arrangements, it next must decide on staffing assignments. Once the organization is set up, it is important to develop the organizational chart, which is a matter of taking each of the job assignments and assigning responsibility and reporting lines. Organizations will generally use similar maintenance organizational charts, with few modifications, depending on the size and development of the maintenance staff. Figures 2-7–2-10 are a sample series of maintenance organizational charts and reporting schemes.

In overall reporting schemes, it is important for the operations, engineering, and maintenance managers to report to the same individual (Fig. 2-7). By all three groups reporting to the same level, the necessary balanced input for the plant management to make decisions is provided. For example, if maintenance needs the equipment for preventive maintenance or needed repairs, and operations wants to continue running the equipment instead of off-loading the production to other equipment, management can hear both sides and make the appropriate decision. If maintenance reports to operations, the plant management may never hear how important the repairs are, and when the equipment does fail, the maintenance organization is blamed unfairly, when the situation was out of their control.

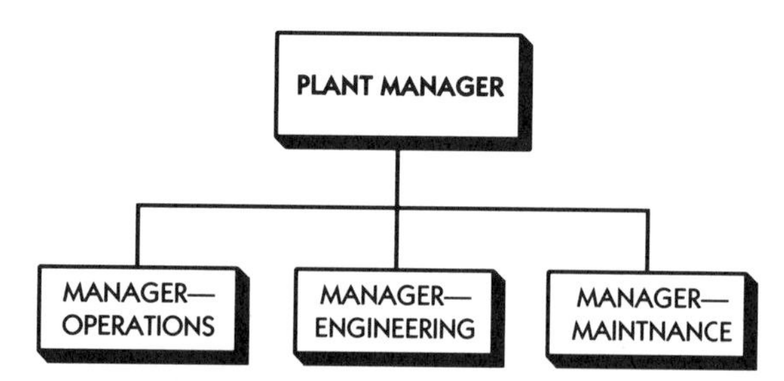

Figure 2-7. Maintenance/plant organizational arrangements.

When maintenance reports to engineering, it is common to find project-type work given importance over preventive and repair maintenance tasks. This situation will reduce the life of the capital equipment, with maintenance receiving the blame. In the most optimum situations, maintenance reports to the plant manager directly.

The basic maintenance organization is pictured in Fig. 2-8. This organization consists of the maintenance manager, the maintenance supervisor, and the planner. In Fig. 2-8, the numbers in each of the boxes are an average of the number of employees that each position should be responsible for. The example given is an organization with 30 "multicrafted" employees. Every supervisor is assigned an average of 10 employees, and each planner is responsible for 15. This arrangement may take coordination, since both planners may be working with the same crew.

A larger organization is pictured in Fig. 2-9, adding a second level of supervision. Now there is a group of line foremen reporting to the maintenance supervisor; there are planners, reporting to the planning supervisor; and there is an engineering coordinator, responsible for the engineering projects for maintenance. All three managers report to the maintenance manager, ensuring proper coordination between the groups.

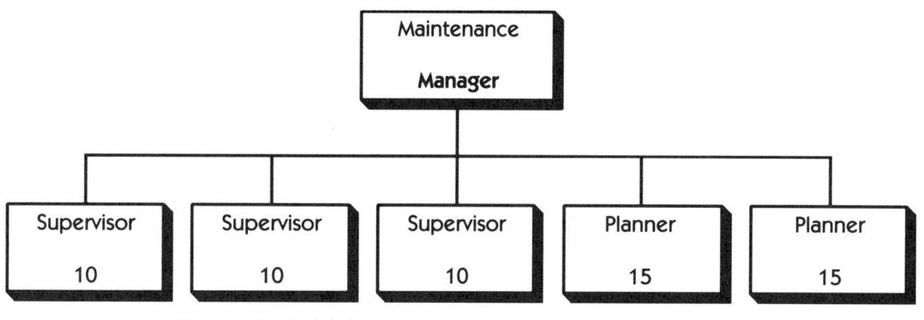

Figure 2-8. Maintenance organizational chart.

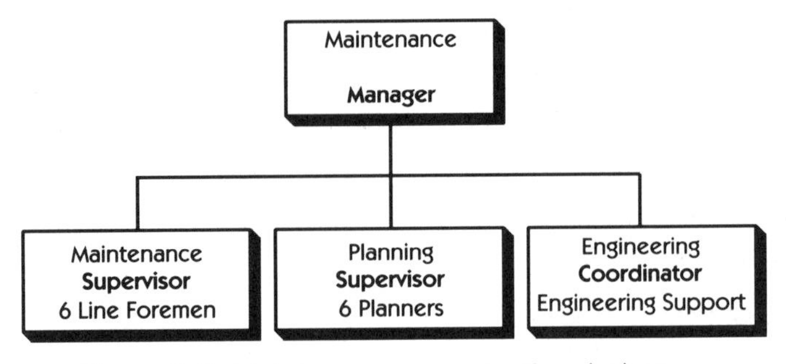

Figure 2-9. Maintenance organizational chart.

Figure 2-10 shows a larger, more craft-oriented organization. The supervisors and planners are organized by craft lines, which is characteristic of companies with strong union environments. Under this arrangement, the planning is more difficult, with the organization, as a whole, inflexible. Also with this arrangement,

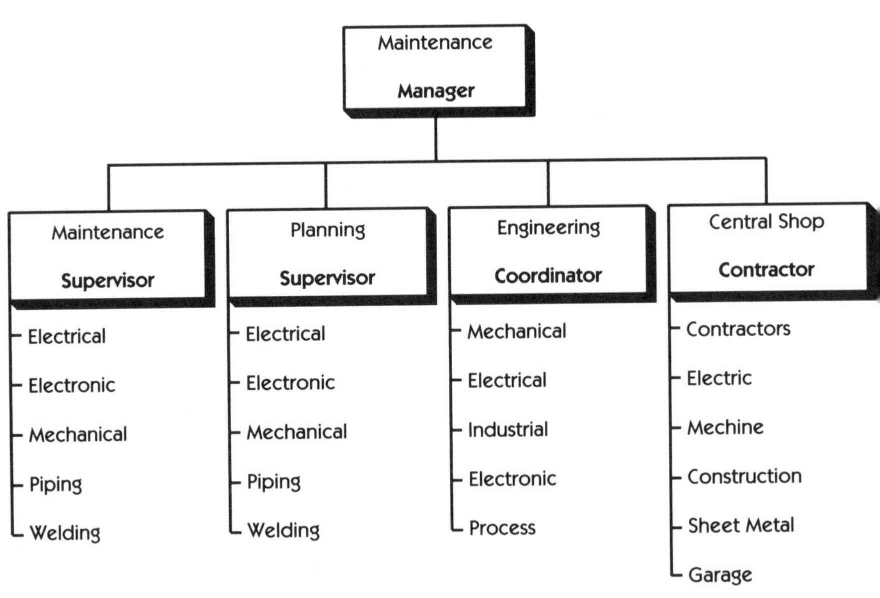

Figure 2-10. Maintenance organizational chart.

the engineering coordinator may have other engineers assigned to his supervision, these being divided into craft groups similar to the planners. The addition to this figure is the coordinator for the central shops. This individual will work with the planner and engineering coordinator to ensure all activities are completed by the time the work is ready to be performed by the craftworkers assigned to the supervisors.

While no organization will use the organizational charts presented exactly as shown, some derivative form will be selected. The important points made in this section are to have an organizational chart, with names, and for it to be current. Maintenance is a hard organization to manage, but without responsibility and accountability, it would be impossible.

In addition to the organizational chart, responsibility and accountability are further defined by detailed job descriptions. Job descriptions do not exist for someone to say "That's not my job!" They exist to help establish responsibility and accountability. If areas are vague and ill-defined, it makes it too easy for critical items to be omitted or overlooked. Job descriptions should be provided for

1. Managers
2. Supervisors
3. Planners
4. Craft lead personnel
5. Craft positions

While the descriptions should be specific, the line "and other duties that may be assigned by management from time to time" removes some of the complaints of narrow-scoped job descriptions.

Attitudes

Since maintenance is not a production or assembly line type of work, attitude plays an important role. How management perceives maintenance, how maintenance perceives its role, and the attitudes of managers and craftworkers toward one another is also important. The attitude management shows toward the maintenance craftworkers helps to establish pride in workmanship. If an organization is in a "fire-fighting" mode, strictly fixing it as quickly as it can be fixed, the workmanship suffers. Craftworkers will get into a habit of fixing it to "just get by." When this environment begins to change to a more proactive environment, the craftworkers have difficulty adjusting to the "fix-it-right-the-first-time" attitude. It is the same problem anyone has when trying to break habits that have become engrained. The in-fighting between maintenance and production/operations/facilities also must end. If maintenance is to contribute to the overall profitability of the corporation, all parts of the organization must be given responsibility and accountability.

Some examples of areas where attitude toward maintenance manifests itself are:

Maintenance shop locations

Maintenance equipment

Maintenance incentive programs

Maintenance repair shops should be located in areas convenient to the job locations. It should be easy for rebuildable items, repairs, etc., to be brought into the shop area, where the larger tools are located. This means adequate clearance for fork lifts, overhead cranes, or other transportation methods. Repair shops should also be located in areas of the plant where excessive noise levels do not make working in the area difficult. For example, in one plant, the maintenance shop was located beside the plant's rock crushers; the noise level made work without hearing protection impossible. In addition, the lathes for the maintenance shop were there also and the vibration made any finishes smoother then 0.010 in. impossible. Needless to say, the maintenance organization did not have a high sense of self-worth.

The equipment in the maintenance shop is also very important. The quality of the tools helps to determine the quality of the work performed by the shop. If the maintenance department does not have the tools and equipment necessary to maintain the plant equipment, it can hardly be said they do not do their jobs. For example, it is difficult to maintain solid-state control equipment with a VOM multimeter instead of an oscilloscope. It is a measure of the importance of the maintenance organization when they are asked to produce precision work with old, worn out tools and equipment.

Maintenance incentive programs are not properly utilized in most plants and facilities to produce motivated craftsmen. Incentive programs can be tied to uptime, production rates, or total departmental operation for the purpose of motivating the maintenance workforce. If the maintenance organization feels that by increased performance they can increase their financial status, they also can work more productively.

Conclusions

The organization for maintenance can be varied and adjusted to fit many circumstances. Some of the options detailed in this chapter are used by organizations around the world. The main points to remember are:

1. All organizations exist to accomplish certain goals or objectives. Maintenance is no different; be sure yours are known and accepted.

2. Organizing the maintenance function is important. Incorrectly organizing the resources can result in excessive maintenance costs.

3. Contractors are being increasingly used in the maintenance environment. Careful policy decisions can make contractors cost effective.

4. Attitudes toward maintenance are shown by the way it is treated when it comes to dedicating resources. Always ensure that maintenance has proper tools, proper locations, and incentives to work.

3 Maintenance Training

Training has been called one of the biggest weakness of the present maintenance structure in the United States. It has been estimated that a company should spend approximately $1,200.00 per year for training of maintenance supervisors. Companies should spend approximately $1,000.00 per year for each craftworker. (See Fig. 3-1.) In fact, if you do not provide some training for a craftworker in an 18 month time period, his skills become dated.

In self-examination, when was the last formal training program for your craftworkers? For your superivisors? For your planners? The importance of training cannot be overstated. Without good quality training programs, a maintenance organization will never be cost effective. We will examine the three areas just mentioned and explore some alternative methods of training.

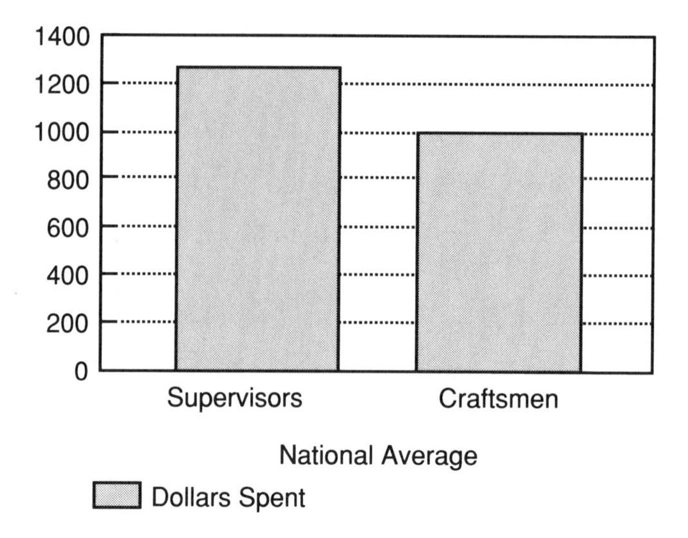

Figure 3-1. Average training dollars required.

Craft Training Programs

The first level of training must be the apprentice training program. This level of training takes the "man on the street" and gives him the training necessary to become a skilled tradesperson. The training program must be a combination of on-the-job training coupled with classroom training. Most good programs will be three to four years in length, with hands-on lab sessions used along with the classroom settings.

Some companies work with local vocational schools to fill entry level craft positions. This allows the company to specify some of the material that must be covered in the program. The vocational school benefits because of the assistance in placing the students when they complete the program. Another option is university level training, but this is generally used for more advanced training later in a craftworker's career.

A second option for the craftworker entry level program is correspondence-type programs, which is completed by craft-people at their own pace. This program is one of the most economical methods used in maintenance training. Closely related to this are the "canned" programs that contain information related to each craft line. The disadvantage to the correspondence program is the material is generic, dealing with theory, not real world experience. While this is good for engineering level training, it is not satisfactory for craft level training. A second problem relates to the lack of a knowledgeable instructor to help the apprentice understand and apply the material being studied.

While the objection about the material being generic is true in most cases, there are at least two vendors who have programs that are specific in dealing with the repair and troubleshooting of components. These programs have been successful and are commended by craftspeople as beneficial. In fact, some of the journeymen, after they saw the quality of the material the apprentices were using, went out and bought their own set of materials for reference.

Another training option is having materials developed inside the company or by an outside firm specializing in maintenance training materials. These materials are developed from a job needs analysis. This is a study that determines just what an individual needs to perform their specific job properly. The materials are then written and illustrated from this needs analysis. The materials are accepted readily, since all the material apply specifically to the craftworker's job. The largest obstacle to using this type of material is the cost. It is expensive to pay to have the needs analysis performed, the materials written, the artwork done, and the materials printed and bound. However, the advantage of having a job-specific training program that the apprentice can take at their own pace can be worth the cost of this program.

The best option (also the most expensive) is to set up your own craft training center; it should have the classroom, lab, and real world setting necessary to produce qualified, competent craftworkers. One such program in existence in the early 1970s became a corporate model for one steelmaker. Of course, the hard times for steel in the mid 1970s put an end to it, since training (especially for maintenance people) was viewed as an unnecessary expense. It has since been restarted, but not on the scale it had in its prime. However, the program structure was designed for a "super-craft" environment, requiring proficiency in mechanical, electrical, and fluid power. The students were required to complete 1040 hours of classroom training, in addition to 3 years of on-the-job training.

The program took different forms depending on the economic situation, with the apprentices attending class on a 1 day per week format; on a 1 week per month format; and on a month on month off format. The optimum appeared to be the 1 week per month format. This allowed the instructors to teach 3 weeks out of the month with 1 week just for preparation of the next cycle.

The five classrooms were set up with heavy tables (72 in. × 30 in.) where the students sat two to a table. The metal top tables were necessary when the instructors would bring a display or lab project into the classroom. There were five labs, three for mechanical, one for fluid power, and two for electrical (one DC control, one AC control). The curriculum consisted of the following:

260 hours DC electricity

260 hours AC electricity

260 hours mechanics

260 hours fluid power

The course outlines were not theory intensive, concentrating instead on the maintenance and troubleshooting of the components and circuits. The apprentices received just enough design theory to understand why certain designs were used, but not enough to do design. For example, in the mechanics class dealing with V-belts, there was enough theory to understand speed differentials and forces generated by the rotation. The thrust of the discussion was how these forces create problems in maintaining the V-belts; why the tension requirements are critical; why alignment is important; why the bearing adjustment is important; why the condition of the sheaves is important; etc. These points made the material applicable to their jobs.

This brings us to the single most important factor in the program: the instructors. The instructors would spell success or failure of the program. *If* the instructor could not relate the theory to the "real world" needs of the apprentices, the course would be a waste of time. For this reason the instructors were journeyman craftworkers with experience. They were required to have good presentation skills, to be able to develop logical, coherent outlines, and to develop tests for their materials. Without use of instructors who were respected for their job skills by their peers, the program would have been a failure.

The training must be conducted in conjunction with on-the-job training also. This was the hardest part of the program; ensuring that if the apprentice was clearing hydraulic circuits, that they had a chance to work on some hydraulic circuits that month. Good communication with the apprentice's department heads and supervisors generally would help in coordination.

The reason for presenting this program was not to paint a pie-in-the-sky picture for anyone, but rather to show the importance that some companies put on craft training. *If* this effort to obtain a fully trained work force is not made, do not expect to achieve a high level of skilled service from your maintenance workforce.

Journeyman Training

Journeyman training is usually related to specific tasks or equipment maintenance procedures. Journeyman training courses can be conducted by in-house experts, vendor specialists, or outside consultants. The training may address a new technology that is being brought into the plant. For example, when vibration analysis was first being introduced into the maintenance environment, there were extensive training programs in the use of vibration analyzers offered by the vendors and consultants. These programs were addressed to the journeyman level craftworkers that would be involved in the programs.

When new equipment is purchased and installed in the plant, training programs are performed by the vendor on the care and maintenance and troubleshooting of the equipment. Again, it is the journeyman level craftworkers involved in the programs.

Good journeyman craft training programs should be a part of any complete maintenance training program.

Cross Training or Pay for Knowledge

This subject is included in the training section, since it is becoming increasingly common in progressive organizations. It is a sensitive subject, since it generally involves negotiations with the union representation. This type of program is essential if the U.S. maintenance costs are going to be brought in line with the costs incurred by overseas competition.

The cost savings for the company is found in planning and scheduling the maintenance activities. For example, consider a pump motor change out. In a strict union environment, it would take:

A pipe fitter to disconnect the piping

An electrician to unwire the motor

A millwright to remove the motor

A utility person to move the motor to the repair area

The installation would go as follows:

A utility person to bring the motor to the job area

A millwright to install the motor

An electrician to wire the motor

A machinist to align the motor

A pipe fitter to connect the piping

As can be seen, not only are many people involved, but the coordination to ensure that all crafts are available when needed without delay will become extremely difficult. In a "multiskilled" or "cross trained" environment, there would be one or possibly two craftworkers sent to the job to complete all the job tasks. The advantages of costs and coordination are obvious.

But what are the advantages for the employees? First, the action helps to ensure maintenance is a profit center. This means being cost effective, but also running the department as a business. For example, could you picture the following production scenario:

One person transports material to the job site

One person inserts the part in the machine

One person drills the hole

One person finishes the hole to specifications

One person machines the finish

One person takes the part out of the machine

One person moves the part to the next process

This would not be a well managed, cost effective operation, would it? Should maintenance be any different? Shouldn't the goal be the same: to maximize the utilization of all assigned resources? Shouldn't the employees work to contribute to the profitability of the corporation? Our overseas competitor's employees do.

A second advantage is that the cross trained employees have

additional skills which ensure they are valuable contributors to the corporation's goals. This increases their self-esteem and value. But most of all, any cross training effort must have financial rewards to the employees. Since cross trained employees are more valuable to the company, they should receive a high rate of pay. For example:

Apprentice—pay level 1

Journeyman (one craft)—pay level 2

Journeyman (one craft, apprentice one craft)—pay level 3

Journeyman (one craft, apprentice two crafts)—pay level 4

Journeyman (two crafts, apprentice one craft)—pay level 5

And the list could go on, depending on the crafts involved.

These types of programs are growing increasingly popular in the United States, as they must. If we are to take advantage of potential cost savings, cross training is one of the most important areas to investigate.

Planner Training

Maintenance planners should come from craftworkers who have good logistics aptitudes. However, planners need training beyond the skills required by craftworkers. They need programs teaching some of the following subject areas:

Maintenance priorities

Maintenance reporting

Project management

Inventory management

Scheduling techniques

Computer basics

If such training is not provided, it will be difficult to achieve the level of proficiency necessary to have a successful planning and scheduling program. Training is one of the most important factors in the development of a good maintenance planner.

Where does one get the training materials necessary for a planner training program? The sources are similar to maintenance craft training materials:

Correspondence courses

University sponsored seminars

Public seminars

Maintenance consultants

Maintenance software vendors

The content of the planner training program will vary depending on whether the organization is a facility, process industry, food service, vehicle fleet, etc. Each will have its own uniqueness, but many of the basic principles will be the same. Good, competent, skilled maintenance planners will pay dividends for the investment in their training many times over.

Supervisor Training

Front line maintenance supervisor positions are filled 70% of the time from craft or planner promotions, so that they will be familiar with the assignments which they will now be responsible to supervise. In a personal observation, it is rare when an engineer or another staff person can make the transition to becoming a front line maintenance supervisor. The lack of craft knowledge or experience cannot be compensated for in actual job supervision.

One of the major mistakes in making front line maintenance supervisors is one week they are in the craft group, the next week they are the supervisor of that group. In many cases, it is sad to say, there is no training given. It is almost as if management wants them to pick up their management skills by osmosis. Good supervisor training programs should be implemented before supervisory responsibilities are assumed. Some areas that should be addressed in these programs are

Time management

Project management

Maintenance management

Management by objectives

The support and understanding of the front line supervisor will help to determine the success or failure of many of the programs implemented by upper management; they should receive the training necessary to ensure that their careers are successful, so that the success carries on to the rest of the organization.

Conclusions

Training is important to all levels of the maintenance organization. Unfortunately, in a volatile financial environment, maintenance programs are the first to be.cut back. This is especially true of programs that are perceived by management as nonessential, which includes, in most organizations, training. This shortsighted management philosophy must change if maintenance is to be managed successfully.

4 Work Order Systems

Work order systems are one of the keys for successful maintenance management. A work order is the document used to collect all necessary maintenance information. Work orders have been described in many different ways, but for the purpose of this text, we will use the following definition:

> A work order is a request that has been screened by a planner who has decided that the work request is necessary and what resources are required to perform the work.

Work orders should not be implemented by just the maintenance department, without regard for other parts of the organization. Figure 4-1 lists the groups that should be involved in the use of a maintenance work order system.

Maintenance, of course, would be the primary user of the work order. Maintenance requires information such as:

What equipment the work needs to be performed on

What resources are required

A description of the work

Priority of the work

Date needed by

There is other information that may be required, depending on the type of facility or plant the work order system is being used in. The main point is the maintenance organization must get the information needed for good management decisions. If the information cannot be obtained from the work order, it is unlikely reliable information will ever be available from another source.

Operations or facilities also need input into the workorder process. They must be able to request work from maintenance easily. If they have to fill out 15 forms in triplicate, chances are they will not participate in the use of the work order, which thus lessens the effectiveness of the work order. The work order system, whether manual or computerized, must be easy for the operations/facilities personnel to use. They should only be required to fill in brief information, such as

Work orders should satisfy and be used by

* Maintenance

* Operations/Facilities

* Engineering

* Inventory/Purchasing

* Accounting

* Upper Management

Figure 4-1

Equipment work is requested on

Brief description of the request

Date needed

Requestor

This information can then be used by the planner to complete the work request and convert it to a work order.

Engineering staff also needs input into the work order system, since they are usually charged with the effectiveness of the preventive/predictive maintenance programs. Of course, they will want to be able to request work for engineering services, but they also need access to historical information. The historical files, if accurate and properly maintained, can provide the information needed to operate a cost effective preventive maintenance program. Without accurate information, the preventive maintenance programs become guesswork. So the engineering staff will need information such as

Mean time between failure

Mean time to repair

Cause of failure

Repair type

Corrective action taken

Date of repair

Proper utilization of this information will enable the engineering staff to optimize the preventive maintenance program.

The inventory and purchasing departments will also need information for the work order system, especially from the planned work backlog. If the work is planned properly, the inventory and purchasing personnel will know what parts are needed and when they are needed. Good historical information on maintenance material usage will assist the inventory and purchasing personnel in setting max/min levels, order points, safety stock, etc., of maintenance materials. The information required by inventory and purchasing would include:

Part number

Part description

Quantity required

Date required

Accounting needs information from the work order to charge the right accounts for the labor and materials used to perform the maintenance task. This charge system is handled differently in many locations; however, the following are the most common types of accounting information gathered:

Cost center

Accounting number

Charge account

Departmental charge number

The use of this information on the work is important, if accounting is to cooperate in the use of the work order form.

Upper management is interested in the information that can be gathered from multiple work orders. This means that it is essential to make the information easy to extract from the work order. Summary-type information should be complied from completed work orders, work orders in process, and work orders awaiting scheduling. If the information is not easy to extract from the work order, managers can spend days gathering the information. The use of check boxes for key information fields can be invaluable for this. Computerized systems make this task easier, but only if they are properly designed.

In summary, the objectives of the work order system are listed in Fig. 4-2.

* A method for requesting, assigning, and following up work
* A method of transmitting job instructions
* A method for estimating and accumulating maintenance costs
* A method for collecting the data necessary for producing management reports

Figure 4-2. Work order objectives.

Types of Work Orders

In any work order system, it is necessary to have several types of work orders. The most common are

Planned and scheduled

Standing or blanket

Emergency

Shutdown or outage

Planned and scheduled work orders have been described in brief. These work orders are requested and screened by a planner; then the resources are planned, the work is scheduled, and work information is entered in the completion process and the work order is filed. This work order is the most common type, and it will be discussed in more detail in Chapter 5.

The purposes of standing or blanket work orders are listed in Fig. 4-3. The types of jobs these work orders are written for are the 5–30 minute jobs, such as resetting a circuit breaker or making a quick adjustment. If you were to write a work order for each of these jobs, maintenance would be buried in a mountain of detail, which could not be compiled effectively into meaningful reports. These standing work orders are written against an

* **Repetitive small jobs where the cost of processing the paperwork exceeds the cost of doing the job.**
* **Fixed or routine assignments where it is unnecessary to write a work order each time it is performed.**

Figure 4-3. Standing or blanket work orders applications.

equipment charge or an accounting number. Whenever a small job is performed, it is charged to the work order number, but the work order is not closed out. It remains open for a preset (by management) time period, then it is closed and posted to history. A new standing work order would be opened at that time.

One problem with standing work orders is people feel that they may be used like credit cards for charging time for the craftsmen that is not accounted for. It does happen occasionally; but, when the charges are closed out on the work order, offenders can be spotted. Computerized systems make this accounting much easier, since they can quickly compile a list of all personnel who have charged time to a work order. In addition, some of the more sophisticated systems can even display the percentage of time any craftwork charges to a type of work order. Therefore, if offenders are suspected, it is easy to find them. However, this example is usually the exception, not the rule. Most employees will not abuse a standing work order system.

Emergency or breakdown work orders are generally written after the job is performed. Breakdowns require quick action, and there is usually not enough time to go through the planning and scheduling of the work order. In most cases, the craftworker, the supervisor, or the production supervisor will make out the emergency work order after the job is completed. The format of the emergency work order is similar to the work request, in that only brief, necessary information is required. When the work order is posted to the equipment history, it should be marked as an emergency work order, which allows the analysis of the emergency work orders by

Equipment identification

Equipment type

Department

Requestor

Analysis of the patterns of emergency work can help identify trends, which can be invaluable when planning maintenance activities.

The typical flow of a trouble call or emergency work order is given in Fig. 4-4. The need for a central call-in point is to prevent overlapping of assignments. If the calls are taken at various points, there may be several technicians dispatched to the same job.

When the technician or supervisor arrives at the job site, he may realize that the job is more complicated than the call indicated. If the work required is going to exceed a certain cost or time limit, the job is routed back to the planner for analysis. If it is going to be easier to coordinate and plan by scheduling, the planner takes control of the work order and schedules it as soon as the material and labor resources are available. This allows for

* **Call comes to central receiving point.**
* **Call dispatched to foreman.**
* **If over a certain time or cost limit, planner should analyze the call.**
* **When materials and labor are available, schedule the call.**

Figure 4-4. Trouble calls—typical flow.

cost effective maintenance activities, instead of wasting labor productivity waiting to do a job.

Shutdown or outage work orders are for work that is going to be performed as a project or when the equipment is shut down for an extended period. These jobs are marked as outage or shutdowns and should not appear in the regular craft backlog. This work is still planned, ensuring that the maintenance resource requirements for the shutdown/outage is known and ready before the shutdown/outage begins. This prevents delays and maximizes the productivity of all employees involved. In many cases, the work order information is entered into project management software to do a complete project schedule. Computerized maintenance management software does not include enough features of project management software to make it an acceptable scheduling alternative. However, some vendors have included interfaces to project management systems, which tends to correct this deficiency.

Obstacles to Effective Work Order Systems

Figure 4-5 outlines some of the most common work order problems. These problems will impact the success of a work order system; this impact ranges from to nuisances total ineffectiveness.

* Inadequate preventive and predictive maintenance programs.
* Lack of labor controls.
* Inadequate stores controls.
* Poor planning disciplines.
* Lack of performance controls.
* Inadequate or inaccurate equipment history.

Figure 4-5. Work orders—typical problems.

Preventive/predictive maintenance programs are key elements in operating a work order system. If an organization is in a reactive ("fire-fighting") mode, it has little or no time to operate a work order system. From the previous information it is obvious that it takes time to provide the information necessary to satisfy the work order system. When an organization runs from breakdown to breakdown, there is little or no time to record the information. If the information gets recorded, it is generally sketchy and inaccurate.

When companies are in a proactive mode, with proper preventive/predictive maintenance programs in place, the work is planned on a regular schedule, with 25% or less emergency activities. This situation provides the supervisors and planners with the needed time to utilize the work order reporting system properly.

A lack of controls for the maintenance labor resource is a second factor that prevents optimum usage of a maintenance work order system. The following problems are common with labor resources:

Insufficient personnel of one or all crafts

Insufficient supervision of personnel

Inadequate training of personnel

Lack of accountability for work performed

Without controls in these areas, inadequate or unacceptable resources result when the planner tries to schedule the work. It is important to have the labor resources properly controlled for a work order system to be effective.

A lack of stores controls can reduce the work order system to total ineffectiveness when materials are required. If the planner does not have accurate, timely information concerning the materials in the maintenance stores, he cannot schedule the work. If the craftworker has a work order, requiring certain materials, but the materials are not available when they are needed, the time wasted obtaining the materials lowers the productivity of the craftworker. It is important for the planner, supervisors, and craftworkers to have current information about the stock levels of maintenance inventory items. Most consultants feel that maintenance materials are the most essential part of a good maintenance planning program.

Poor planning disciplines affect the work order system, because most of the information on the work order is not reliable. When this is the situation, work orders fall into disuse, resulting in a discontinuance of the work order information flow. Job plans must be accurate and realistic if the work order system is to be successful. If companies do not have a work order planning system, they really do not have a work order system.

Lack of performance controls is really a lack of follow up on management's part. Once a job plan is produced on a work order, it should always be audited for compliance. This audit may highlight weaknesses in

Planning

Scheduling

Supervision

Craft skills

Any deficiencies can then be corrected. However, if performance controls are not used, the lack of accountability will create an disorganized effort, again allowing the work order to fall into disuse.

Inadequate or inaccurate equipment history hinders the work order system, since none of the information used to make management decisions will be reliable. The managers will not be able to base budget projections, equipment repair forecasts, or labor needs on historical standards.

Conclusions

The work order system is the cornerstone for any successful maintenance organization. If the work order is not used, do not expect much of a return on investment from the maintenance organization. But, work order problems are not all maintenance related. Unless all parts of an organization cooperate and use the system, true maintenance resource optimization will be just a dream.

5 Maintenance Planning and Scheduling Programs

A survey polled maintenance managers as to their top problems. Figure 5-1 shows over 40% of the respondents felt scheduling was their biggest problem. (Since there was more than one response per respondent, the totals are well over 100%.) Maintenance planning and scheduling is one of the most neglected disciplines today. We will explore some of the reasons for the lack of good planning and scheduling as well as some solutions.

Maintenance Planners and Supervisors

One of the major obstacles to maintenance planning and scheduling is management's reluctance to admit that planners are essential to the program. In fact, Fig. 1-3 shows that two-thirds of maintenance organizations in the United States do not even have planners; however, Fig. 5-2 shows a hidden problem. Organizations that have planners, place responsibility for planning for too many craftworkers on them. Planners should be responsible for

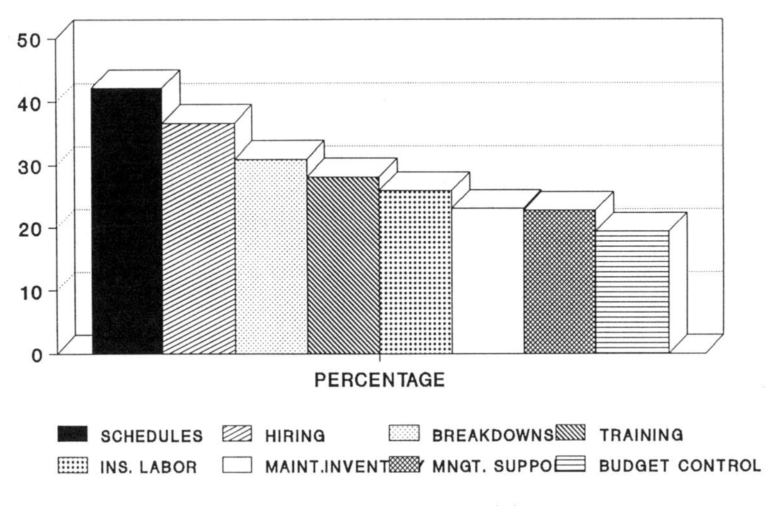

Figure 5-1. Maintenance problems.

15 (optimum) to 25 (absolute maximum) craftworkers. Supervisors, then, are responsible for overseeing the work of an average of 10 craftworkers. Why the difference between the two groups? This can best be answered by examining the job description for the supervisor and planner.

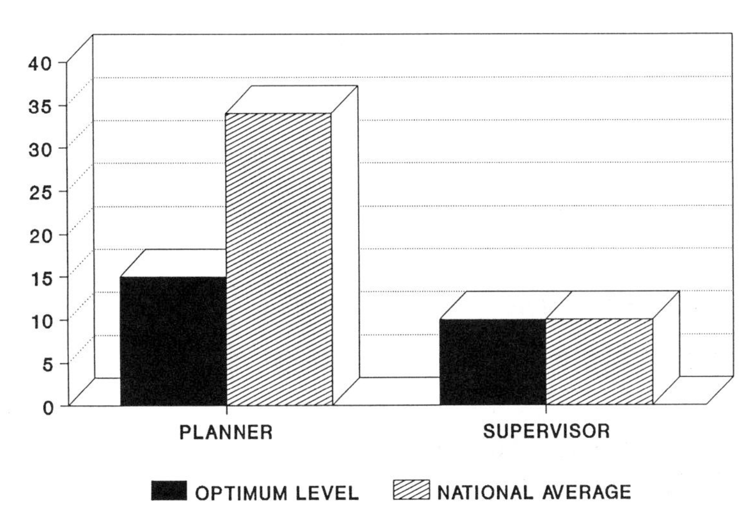

Figure 5-2. Planner/supervisor ratios.

Supervisor's Job Description

Figure 5-3 details the tasks assigned to the typical maintenance supervisor. While some of the duties may vary from area to area, this is a good outline of the responsibilities a supervisor should be charged with. We will now consider them in detail.

Motivating the craft personnel means it is the supervisor's responsibility to see that each of the craft technicians they are assigned to supervise is ready to perform his job each day. This does not mean the supervisor has a whip and chair and drives them to the job. But using good management skills, he sees that they are motivated to perform their assignments. Since the most effective way to motivate is by example, the supervisor would be ready to begin at starting time and would also be available continually to assist the craft technicians during the workday.

When a supervisor determines the craft/skill/crew for the job, they are not determining what craft manpower is required; this has already been done by the planner. The supervisor looks at the job and matches the skill level of the craft technicians to the

* Motivate the craft personnel.
* Determine craft/skill/crew for the job.
* Coordinate and follow-up the job.
* Perform safety and quality monitoring of the job.
* Hiring, firing, and pay reviews of the assigned employees.
* Recommend improvements and cost reductions.
* Identify the causes of failures for breakdowns and repetitive repairs.
* Recommend skill levels and training courses for apprentices and journeymen.

Figure 5-3. Maintenance supervisor's typical job description.

job, or, if it is a larger job, matches the right skill level for several crafts and sends them all out to the job. The supervisor is responsible for determining who works on each job.

The coordination and follow-up of each job indicates the supervisor is in the field with the craft technicians. He does not sit behind a desk, since he cannot see what is going on from there.

The safety and quality of the job again indicate the supervisor must be out with the craft technicians. However, this also indicates the supervisor must have some knowledge of the job skills and techniques, if the quality and safety are his responsibility.

If the supervisor is to be responsible for the hiring, firing, and pay reviews of the employees assigned to his supervision, he must have been trained properly to do so. One large mistake often made here is the maintenance manager wants to do the reviews, discipline, commendations, etc. But, who better knows the work habits and skills of the employee? It would be the direct supervisor, not the next level of management. Proper management training will enable the supervisor to assume this responsibility.

If a supervisor is to recommend improvements and cost reductions, he must be technically competent, understanding the jobs his people are asked to perform and the processes they are asked to maintain, and have a basic understanding of the engineering principles involved in the design of the equipment being maintained. While this appears to be a demand on the technical side of a supervisor, it merely highlights that a maintenance manager must be skilled, if he is to be successful.

Identifying the causes of failures or of repetitive breakdowns highlights the troubleshooting skills of the supervisor. If he is not effective in this area, his people merely become parts changers. The supervisor's feedback to the work order system relating to cause of failure can help make adjustments in the preventive/predictive maintenance program, as well as give engineering some valuable feedback on design flaws.

Recommending the necessary skills and training programs for the craft technicians, again highlights the interaction among the supervisor and his personal. By knowing what they can and cannot do, he is in a position to recommend what new training they need to improve their job knowledge and skills. This improves the attitude of the craft workforce, while enhancing the supervisor's ability to manage.

When examining the supervisor's job description, there are several points that become clear. First, the supervisor must have in-depth job knowledge of the crafts he is going to supervise. This is important because many of the responsibilities listed in the job description require knowledge of the tasks the craft technicians will be performing. Second, the supervisor must be out on the floor with his people. It is sad to say that in the majority of the maintenance organizations today the job of the supervisor has become a "glorified clerical position." This is wrong! A front line maintenance supervisory position should be structured so that the supervisor spends no more that 25% of his time on paperwork. The other 75% of his time should be spent out on the floor with his people.

The reason for this problem has developed over time. Cutbacks, with resulting elimination of clerical support help, have forced the supervisor to assume duties not in the job description. But one worse mistake is management's unwillingness to dedicate maintenance planners to plan and schedule the maintenance activities. Figure 5-4 emphasizes this point. Maintenance depart-

#1 MISTAKE

The first **big mistake** a company will make when initiating a planning program is to try to make the current supervisors plan and supervise.

It does not work !!!

Figure 5-4

ments must be staffed properly, if maintenance supervisors are to fulfill their job assignments. To highlight how important it is to have planners, and to show how their jobs differ from maintenance supervisors, let us look at a planner's job description.

Planner's Job Description

Figure 5-5 highlights the main parts of the planner's job description. We will now explain each of the tasks in more detail.

The planner's job starts when a work request is received. The planner is responsible for reviewing the request and ensures that the request is not currently an active work order. The planner must also clearly understand what the requestor is asking for, so that the work plan he develops will produce the desire results.

If the planner is unclear on what is requested, he will visit the job site. This serves two purposes. First, it ensures that the planner will have clearly in mind what is requested. Second, it allows him time to look for any safety hazards or other potential problems that may need to be documented.

If after visiting the job site, the planner is still unclear on what is being requested, he visits the requestor. This face-to-face discussion will ensure that the work is accurately understood before planning begins.

Once the planner understands the work requested, he estimates what craft groups it will take to do the job and also how long it should take them to do the job. This step is extremely important, for these estimates are the foundation for scheduling accuracy.

The planner next decides what materials are needed. This is where accurate stores information is crucial. Without reliable information about on-hand quantities, maintenance planning will be inaccurate, resulting in maintenance schedules that are unreliable. The planner will ensure that all materials are available, in sufficient quantity before completing this step.

The planner may not find the necessary parts to do the job stocked in the storeroom. It may be necessary to order them

* Reviews requests for work.
* Visits job site for clarification.
* Confers with requestor.
* Estimates the craft labor required.
* Reserves all stores material required.
* Orders all nonstock material.
* Ensures all resources are available before the work order is scheduled.
* Develops standards for repetitive jobs.
* Develops historical job estimates.
* Develops and tracks craft/crew backlogs.
* Determines labor capacity for schedule.
* Prepares weekly schedule for approval.
* Tracks work orders to completion.
* Keeps completed work order file by equipment number.
* Keeps the information showing
    - Date of repair
    - Work order number
    - Accumulated down time
    - Cause code
    - Priority of work
    - Actual labor
    - Actual materials
    - Total cost
    - Year-to-date and life-to-date costs
* Tracks all equipment information including spare parts and manuals.
* Works with engineering to establish and optimize preventive maintenance programs.

Figure 5-5. Maintenance planner's typical job description.

directly from the manufacturer; these are referred to as nonstock items. The delivery date of the nonstock items ordered is the key to further processing of the work order.

The planner will ensure that all the required resources, labor, materials, tools, rental equipment, contractors, etc., are ready before the work is scheduled. This eliminates lost productivity, since everything is ready before the craft technicians begin to work on the job.

Based on prior completions and engineering studies, the planner maintains a file of repetitive jobs. These are jobs that are performed the same way, using the same labor and materials each time. The job is not done on a regularly scheduled basis, but with varying frequencies. He builds a file and statistically averages the actual labor hours and related costs each time the job is done. This figure can become his estimate the next time the job is scheduled, thus increasing the accuracy of the estimates.

Another method the planner may use is keeping the historical file of work orders by equipment. When a job comes up that has been done before, the planner can pull the previous work order from the history file. By copying the job steps, materials, etc., from the previous work order, job planning becomes easier. Of course, the planner would look for completion comments to ensure the previous job plan did not overlook anything.

Since the planner maintains control of the work order file, it is his responsibility to develop the craft backlog. The craft backlog is a total of all the labor requirements for work that is ready to schedule. This figure allows the planner to alert management for the need to add craft labor or to decrease it. The planner will plot the backlog trends for a running 6 month chart, which allows trends to be plotted and tracked.

The planner will also track the labor capacity for each week, so that he can take enough work out of the backlog to make up the weekly schedule. This total will take into account vacations, sickness, overtime, etc. This helps to ensure an accurate schedule will be produced.

By matching work from the backlog to the labor availability, the planner produces a tentative weekly schedule. He presents it to management, who makes any needed changes and approves the schedule for the next week. The schedule is given to the maintenance supervisor at the end of the week, so that he can prepare for the next week. It must be noted that the planner does not tell the supervisor when he will do each job, or who works on each job; that is the supervisor's responsibility. The planner is responsible for weekly schedules; the supervisor for the daily schedules.

When the work orders are completed, the planner receives them, notes any problems, and files the completed work orders in the equipment file. The work order file is kept in equipment sequence for easy access to the equipment repair history.

Each work order will contain the following information:

Date of repair

Work order number

Accumulated downtime

Cause code

Priority of work

Actual labor

Actual materials

Total cost

Year-to-date costs

Life-to-date costs

This information can be complied by management into reports for future decision making. This is a cumbersome task annually, but computerized systems can do this with relative ease.

The planner is also responsible for maintaining the equipment information, such as drawings, spare parts listings, equipment manuals, etc. This is for reference of the entire maintenance organization, but particularly for his own, since he will need this information for planning.

Since the planner has access to the work order files, he can work with engineering to spot any excesses or deficiencies in the preventive/predictive maintenance systems.

As can be seen by reviewing the previous material, the planner has a full time job. The planner differs from the supervisor in that he is more paperwork oriented. The planner should expect to spend 75% of his time on paperwork/computerwork and only about 25% of his time out on the floor, looking over equipment, parts, or spares. This is why planners and supervisors cannot be the same people. They both have full time jobs, with heavy responsibilities.

What kind of job skills should a planner have? These are listed in Fig. 5-6. First and most important, the planner must have good craft skills. If the planner is to be effective in planning the job, he must know how to do the job himself. If job plans are not realistic and accurate, the program will not be accepted by the craft technicians, since poor planning increases their work and causes frustration.

* Must have good craft skills.
* Good communication skills.
* Good aptitude or paper/computer work.
* Clearly understands instructions.
* Good sketching ability.
* Must understand how maintenance functions within the organization's structure.

Figure 5-6. Planner's skill requirements.

The planner must also have good communications skills. The planner is required to interface with many levels of management, operations/facilities, and engineering. Poor communication skills can dramatically impact the relationship maintenance has with any or all of these groups.

The planner must also have a good aptitude for computer/paperwork, since 75% of the planner's time is spent in this type of activity. Some craft technicians cannot make this transition. It is only fair to specify the requirements to them before they become planners.

The planner must also have the ability to understand instructions clearly. In communication with the requestors of work, instructions will be conveyed. If they are not understood, how can the planner transmit them to the craft technicians?

The thought of good sketching ability may seem superfluous, but consider the following scenario. The planner is asked exactly what part needs to be changed. He picks up a pencil and piece of paper and begins to draw the part. It happens all the time, and sketching is an indispensable skill for a planner to have.

Most of all, the planner must be educated to the priorities and management philosophy for the organization. Without a clear understanding, it will be difficult for the planner to function in a manner satisfactorily to him or management. With a good understanding, it can enhance the entire planning program's performance and can contribute to overall acceptance of the program.

Reasons for Planning Program Failures

Figure 5-7 highlights the most common reasons for planning failures. Examining them in more detail can help prevent future programs from meeting with failure. One reason programs fail is the fact that job descriptions are not clear and there are overlapping job responsibilities. This means the first planner thinks the second planner is doing it. And, of course, the second plan-

* More than one person was responsible ...
  Something was overlooked.
* Planner was not qualified.
* Planner was careless.
* Planner did not have enough time to plan properly.

Figure 5-7. Reasons for planning failures.

ner thinks the first planner is doing it. In fact, no one plans the job and the planning program loses credibility. Elimination of this problem involves strict planning lines. Whether the plans are by craft, crew, department, supervisors, etc., make the responsibilities clear, and then monitor them. This problem can be eliminated with good management controls in place.

Planners who are not qualified will quickly end a planning program. Unrealistic or ridiculous job plans will destroy the credibility of the planning program, quickly putting an end to the entire program. The first requirement of having the proper job skills must be met, followed by the rest of the qualifications. Planners should be trained properly and given the opportunity to apply the training. However, for the sake of the whole program, an ineffective planner must be removed.

Planners can get careless, and job plans will suffer. When this becomes a problem, proper disciplinary procedures should be implemented. However, before these procedures are implemented, check to ensure that management is not to blame. It is quite common for carelessness to be confused with the last item.

There is usually one reason why planners do not have enough time to plan properly: the ratio of planners to technicians is not correct. As discussed earlier, the ratio should be 1 : 15 at the optimum. A ratio of 1 : 25 could possibly be used if the working conditions and type of work planned are optimum. Anything above 1 : 25 spells certain failure for the program.

Consider the steps involved in planning a work order that

were discussed previously. Could any one person do that for 25 work orders per day? How about 50 work orders per day? Consider how many work orders each craft technician completes each day. Multiply that times the number of technicians per planner and you have the work load for the planner. Overloading the planner is the most common problem. Elimination of this problem helps to ensure a successful planning program.

Benefits of Planning

What type of benefits do we see to planning? First of all, there is the cost savings. The following are some documented savings companies have had using planned maintenance versus breakdown or emergency maintenance:

	Planned	Unplanned	Savings
Job 1	$30,000.00	$500,000.00	$470,000.00
Job 2	$46,000.00	$118,000.00	$72,000.00
Job 3	$6,000.00	$60,000.00	$54,000.00

These were identical jobs that were performed once in a breakdown mode, and a second time in a planned and scheduled mode. The cost savings from these three examples would pay for a comprehensive planning program for a long time.

In addition to the bottom line dollar savings, there is an increase in maintenance productivity, which also affects the morale of the workforce. Consider the definition of hands-on time in Fig. 5-8. The national average for hands-on time is less than 50% for maintenance technicians. In some reactive organizations, it is below 25%. Why is this happening? Figure 5-9 highlights the productivity losses resulting from unplanned work. These delays and losses are so common, they need no elaboration.

In defense of the craft technicians, they are a skilled craft group. They, like any other craftsmen, want to do the best job possible. Lack of cooperation and coordination on management's

Hands-on time
is the time when the craftsmen physically have
their hands on the job
that is in progress. This is the time
we are really paying for!!!!

Figure 5-8. Hands-on time definition.

part, impacts their ability to do so. Figure 5-10 lists what is needed to ensure the crafts ability to take pride in their work activities. Again, without elaboration, all of these elements are part of a good planning and scheduling program.

Planning is an integral part of any successful maintenance organization. Planning affects everything from the bottom line to craft morale. If you have tried planning in the past and failed, consider some of the suggestions for success and try them. If you are planning now, review this section and try to optimize the planning.

If you have not tried planning maintenance work, or are unwilling to try it, you can stop reading this chapter now and move on to the next one. Without good, accurate job plans, scheduling maintenance work is impossible. Reading the next section with-

* Waiting for instructions and parts.
* Looking for supervisors.
* Checking out the job.
* Multiple trips to the stores and job site.
* No special tools.
* Waiting for approval.
* Too many craftworkers per job.

Figure 5-9. Labor productivity losses.

* They have job instructions.
* They know why the work is being done.
* The materials are available.
* The tools are available.
* Operations/facilities cooperate.
* They are allowed to complete the work assignment.

Figure 5-10. Requirements for crafts to take pride in their work.

out a planning program is an exercise in futility. If you want to plan, are planning, or need further convincing to plan, the scheduling section will be invaluable.

Maintenance Scheduling

In its most simple meaning, maintenance scheduling is the matching of maintenance labor and materials resources to the requests for the maintenance labor and material resources. However, if it were that simple, maintenance scheduling would not be listed as one of the major problems for maintenance managers. The flow of scheduling starts with good job plans, statusing the work order, scheduling the work when resources are available, completing the work when scheduled.

When planning the work order, the planner needs to track the work order through various statuses. Figure 5-11 lists some of the common status codes for work orders. A planner would want to ensure that the work order had cleared all wait codes before the work was statused "ready to schedule." For reasons previously discussed, scheduling work before it can be started decreases maintenance productivity.

The next step is to determine the available labor capacity for

Waiting Codes	Work Codes
Authorization	Ready for schedule
Planning	In process
Engineering	Completed
Material	Cancelled
Shutdown	

Figure 5-11. Work order status codes.

the scheduling period. The most accurate formula for determining maintenance labor capacity is shown in Fig. 5-12. The true labor capacity can be compared to a payroll check. It has a gross amount and a net amount. The gross is the hours worked times the payrate; the net is what is left after taxes, social security, etc. Labor capacity also has a gross amount, total employees times the hours scheduled, plus overtime and/or contract workers. However, you can never expect that much work to be done, anymore than you can expect to spend the gross amount on the paycheck. There are deductions from the labor capacity, such as unsched-

Total gross capacity
Total men × Total hours worked
+ Overtime and/or contract labor
minus
Unscheduled emergencies . . . weekly average
+
Absenteeism . . . weekly average
+
Allotment for preventive maintenance
+
Allotment for standing or routine work . . .
use weekly average

Figure 5-12. Maintenance labor capacity.

uled emergencies, absenteeism, allotments for preventive maintenance work or routine work.

Good scheduling also necessitates knowing the amount of work still to be performed by each craft; this is commonly called the craft backlog. The craft backlog can be accurately determined only if "real world" figures are used. The formula for accurately measuring the craft backlog in weeks is pictured in Fig. 5-13. It should be noted that accurate backlogs involve the open work orders that are ready to schedule, not work that is waiting on something before it can be scheduled. Dividing this total hour figure by the crafts capacity for the week (determined by Fig. 5-12) gives the craft backlog in weeks.

Knowing the craft backlog in weeks helps to determine the staffing requirements for the craft group. A good backlog is 2–4 weeks worth of work. Some companies will allow a 2–8 week range, but beyond that, requestors tend to bump up the priority and circumvent the scheduling process. A craft backlog higher than 4 weeks indicates a need for increased labor. This can be accomplished by

Working overtime

Increasing contract labor

Transferring employees

Hiring employees

Maintenance staffing levels should be determined by craft backlogs:

$$\text{Craft backlog (in weeks)} = \frac{\text{Open work orders ready to schedule (total hours)}}{\text{Craft capacity (weekly)}}$$

Figure 5-13. Maintenance staffing.

A craft backlog lower than 2 weeks indicates a need for reduced labor. The reduction may come in the following ways:

Eliminating overtime

Decreasing contractors

Transferring employees

Laying off employees

To properly manage the workforce, it is necessary to trend the backlog over a time period. This helps to identify developing problems and to evaluate attempted solutions. A good graph should be for a rolling 12 month time period. A sample graph is pictured in Fig. 5-14.

With the above information, the planner is now ready to begin making out the schedule. It should be noted here that the planner is primarily concerned with a weekly schedule. This gives the maximum flexibility for handling unexpected delays

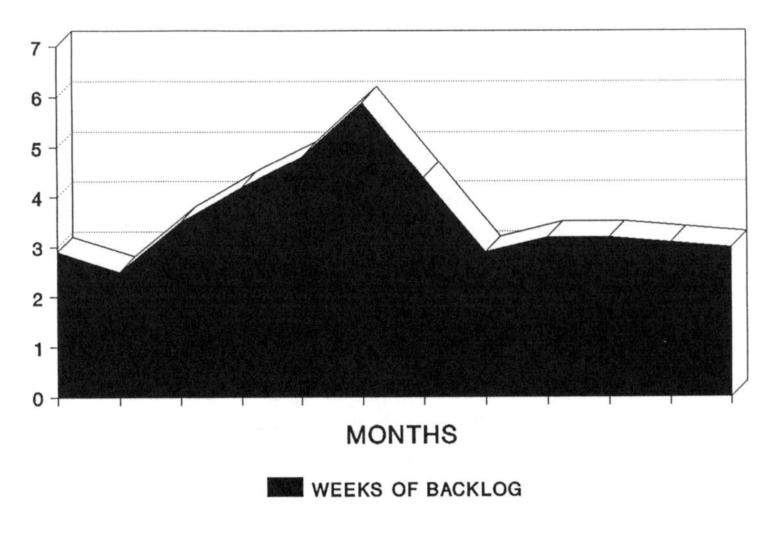

Figure 5-14. Craft backlog.

* Work priority.
* Work already in progress.
* Emergency and breakdown work.
* Standing and minor work.
* Preventive maintenance work—due and overdue.
* Actual craft labor available (absentee, vacations, overtime, contract).
* Craft backlog.

Figure 5-15. Maintenance work order scheduling considerations.

caused by emergency/breakdown work, weather delays, production rush orders, etc. As the planner begins the process, he needs to be aware of the items listed in Fig. 5-15.

Maintenance priorities are decided upon by a variety of features. It is generally the measure of the criticality or importance of the work. Figure 5-16 lists some of the common priorities for a simplified priority system. It should be noted that the simpler the priority system, the more widely accepted it will be. The more complex it is, fewer people will understand it, and fewer people will use it properly. Figure 5-17 shows one of the more

* Emergency or breakdown.
* Urgent, critical (24–48 hours).
* Normal, plan, and schedule.
* Shutdown, outage, rebuild.
* Preventive maintenance.
* Safety.
* Fabrication.

Figure 5-16. Maintenance priorities.

Work Priority Class (WPC)	Production Machine Code (PMC)
10. CRITICAL	10. CRITICAL
9.	9.
8.	8.
7.	7.
6.	6.
5. MODERATE	5. MODERATE
4.	4.
3.	3.
2.	2.
1. UNIMPORTANT	1. UNIMPORTANT

WPC × PMC = Final priority

Figure 5-17. Multiplier priorities: equipment criticality.

complex priority system available. This system allows for input from maintenance and operations as to the importance of the equipment and the requested work. When the two factors are multiplied together, the final priority is derived. The higher the priority, the faster the work gets done. Some systems even allow for an aging factor, which raises the priority so many points for each week the work order is in the backlog. This prevents "lifers" or work orders that never get done.

Referring again to Fig. 5-15, the planner uses the status of the

* Should be 80–90% scheduled.
* Should be planned by experienced technicians.
* Should be processed as backlog—weekly schedule, then daily work.
* Must be flexible enough to accommodate emergency work.
* Should not be scheduled until ready.

Figure 5-18. Maintenance scheduling.

work order to begin the listing. All work already in process should be scheduled first, to eliminate jobs that are partially completed in the backlog. These would be sorted within this status by priority. The next status would be those work orders previously scheduled but not started. These also would be sorted by priority within the status. The next would come the work that is ready to schedule, again sorted by priority. The listing could look as follows:

Work Order Number	Status	Priority	Date Needed
101	In process	10	
102	In process	9	10/21/90
103	In process	9	10/30/90
104	Scheduled	10	
105	Scheduled	8	
106	Scheduled	5	
107	Ready to sched	9	
108	Ready to sched	6	
109	Ready to sched	3	

The planner would deduct the hours required to do each work order from the available capacity for the craft group. When he runs out of hours of crafts labor available, that is all the work he can expect the crew to get done for the next week. Any additional work orders on the list would go into the backlog.

* **Good, accurate estimates.**
* **Good work order system including job instructions, crafts required, required date.**
* **Accurate craft availability.**
* **Accurate stores information.**
* **Accurate contractor information.**

Figure 5-19. Requirements for scheduling.

The planner will now take the schedule to a management meeting. He will present what is scheduled for the next week. The maintenance manager, production/facilities manager, or the engineering manager may request some changes. The planner would make the changes, perhaps deferring some work orders in favor of getting some other work done. Once the schedule has been agreed to, the planner will finalize it and distribute copies to all parties involved, usually on the Friday of the preceding week. This ensures that there is agreement before the week starts.

Conclusions

In conclusion, Fig. 5-18 highlights some requirements for maintenance work and maintenance scheduling. The cooperation among the various groups involved will ensure that these goals are not just "wishes" but will be realities. By following the guidelines here, it will be easy to be successful in scheduling maintenance work.

Figure 5-19 shows, in review, what is required from the planner if the schedule is to be successful. By the planner ensuring that the listed information is correct, accurate scheduling of maintenance activities will be a reality.

6 Preventive Maintenance

Preventive maintenance has become a term with a broad definition. If you ask a room of 20 people to write their definition of preventive maintenance, you will get 20 different answers. In this chapter we will give a generic definition for preventive maintenance, discuss the types of preventive maintenance, and how to implement a good preventive maintenance program.

What is preventive maintenance? Preventive maintenance is any planned maintenance activity that is designed to improve equipment life and avoid any unplanned maintenance activity. In its simplest form it can be compared to the service schedule for an automobile. There are certain tasks scheduled at varying frequencies, all designed to keep the automobile from experiencing any unexpected breakdowns. Preventive maintenance for equipment is no different.

Why is preventive maintenance important? The reasons a preventive maintenance program is necessary are listed in Fig. 6-1. Increased automation in industry requires preventive mainte-

* Increased automation.
* Just-in-time manufacturing.
* Business loss due to production delays.
* Reduction of equipment redundancies.
* Reduction of insurance inventories.
* Cell dependencies.
* Longer equipment life.
* Minimize energy consumption.
* Produce higher quality product.
* Need for more organized, planned environment.

Figure 6-1. Reasons for preventive maintenance.

nance. The more automated the equipment, the more components that could fail and cause the entire piece of equipment to be taken out of service. Routine services and adjustments can keep the automated equipment in the proper condition to provide uninterrupted service.

Just-in-time manufacturing (JIT) is becoming more common in the United States. JIT requires that the materials being produced into finished goods arrive at each step of the process just in time to be processed. JIT eliminates unwanted and unnecessary inventory. JIT also requires high equipment availability. This means that the equipment must be ready to operate when a production demand is made and not break down during the operating cycle. Without the buffer inventories (and high costs) traditionally found in U.S. processes, preventive maintenance is necessary to prevent equipment downtime.

If equipment does fail during an operational cycle, there will be delays in making the product and delivering it to the customer. In these days of intense competition, delays in delivery can result in a lost customer. Preventive maintenance is required so that the equipment is reliable enough to ensure a production schedule that is dependable enough to give a customer firm delivery dates.

In many cases, when equipment is not reliable enough to schedule to capacity, companies will purchase another, identical piece of equipment as a back-up or spare, in case the first one breaks down on a critical order. With the price of equipment today, this can be an expensive solution to a common problem. Unexpected equipment failures can be reduced, if not almost eliminated, by a good preventive maintenance program. With equipment availability at its highest possible level, redundant equipment will not be required.

Reducing insurance inventories has an impact on maintenance and operations. Maintenance carries many spare parts just in case the equipment breaks down. Operations carries additional in-process inventory for the same reason. Good preventive maintenance programs allow the maintenance departments to know the condition of the equipment and prevent breakdowns. The savings from reduction (in some cases, elimination) of insurance inventories can finance the entire preventive maintenance program.

In manufacturing and process operations, each production process is dependent on the previous process. In many manufacturing companies, the processes are divided into cells. Each cell is viewed as a separate process or operation. Each cell is dependent on the previous cell for the necessary materials to process. If the cell has an uptime of 97%, this might be acceptable for a stand alone cell. But if 10 cells, each with a 97% uptime are tied together to form a manufacturing process, the total uptime for the process would be 71% (see Fig. 6-2). This is unacceptable in any process. Preventive maintenance must be used to raise the uptime to even higher levels.

Longer equipment life is a result of performing the needed

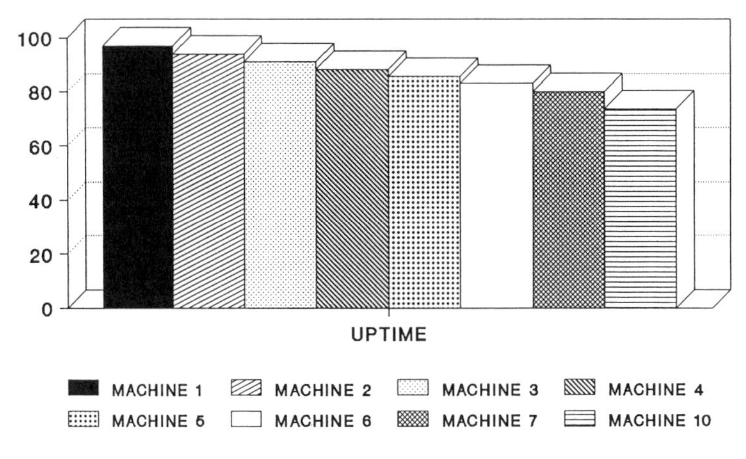

Figure 6-2. Cell uptime for 10 machines.

services on the equipment when required. Using the example of an automobile again, if it is serviced at prescribed intervals, it will deliver a long and useful life. However, if maintenance is neglected, it will have a shorter useful life. Since some industrial equipment is even more complex than the newer computerized automobiles, service requirements may be extensive and critical. Preventive maintenance programs allow for these requirements to be met, reducing the amount of emergency or breakdown work the maintenance organization is required to perform.

Preventive maintenance reduces the energy consumption for the equipment to its lowest possible level. Well-serviced equipment will require less energy to operate, since all bearings, mechanical drives, and shaft alignments will receive timely attention. By reducing these drains on the energy used by a piece of equipment, overall energy usage in a plant could be reduced by 5%, which is certainly another cost reduction to help justify a good preventive maintenance program.

Higher product quality is a direct result of a good preventive maintenance program. Out of tolerance equipment never produced a quality product. World class manufacturing experts realize and emphasize that rigid, disciplined preventive maintenance programs produce high-quality products. To achieve the quality required to compete in the world markets today, preventive maintenance programs will be required.

If operations or facilities were organized and operated the way the majority of maintenance organizations are, we would never get any products or services when we needed them. An attitude change is necessary to give maintenance the priority it needs. This may also involve a management viewpoint change. U.S. management tends to sacrifice long-term planning for short-term returns. This type of attitude is what causes problems for maintenance organizations. The attitude leads to reactive maintenance with little or no controls. When maintenance is given proper attention, it can become a profit center, producing positive, bottom line improvements to the company.

One important point should be made before types of preventive maintenance programs are discussed.

No preventive maintenance program will be truly successful without strong support from upper management of the plant or facility.

There are enough decisions that must be made by plant management in the area of deciding to allow time to perform maintenance on the equipment instead of running it wide open, that without the commitment to the program preventive maintenance will never be performed. Or it will be too little, too late. This support is the cornerstone for any preventive maintenance program.

Types of Preventive Maintenance

The various types of preventive maintenance are listed in Fig. 6-3. A good preventive maintenance program will incorporate all of these types, with the emphasis varying from industry to industry and facility to facility. This list also provides a progressive,

* Routine—lubrication, cleaning, inspections, etc.
* Proactive replacements.
* Scheduled refurbishings.
* Predictive maintenance.
* Condition-based maintenance.
* Reliability engineering.

Figure 6-3. Types of preventive maintenance.

step-by-step method of implementing a comprehensive preventive maintenance program.

Routine maintenance, such as lubrication, cleaning, and inspections, is the first step in beginning a preventive maintenance program. These service steps take care of small problems before they cause equipment outages. The inspections may reveal deterioration that can be repaired through the normal planned and scheduled work order system. One problem develops in companies that have this type of program: they stop here, thinking this constitutes a preventive maintenance program. However, this is only a start; there is more a company can do.

Proactive replacements is a replacement of deteriorating or defective components before they can fail. This scheduling of the repairs eliminates the high costs related to a breakdown. These components are usually found during the inspection of routine service. One caution should be noted: replacement should only be for components in danger of failure. Excessive replacement of components "thought" to be defective can inflate the cost of the preventive maintenance program. Only known defective or "soon to fail" components should be changed.

Scheduled refurbishings are generally found in utility companies, in continuous process-type industries, or in cyclic facilities usage, such as colleges or school systems. During the shutdown or outage, all known or suspected defective components are changed out. The equipment or facility is restored to a condition where it should operate relatively troublefree until the next outage. These projects are scheduled using a project management type of software, allowing the company to have a time line for starting and completing the entire project. All resource needs are known in advance, with the entire project being planned.

Predictive maintenance is a more advanced form of routine inspections. Using the technology presently available, inspections can be performed that detail the condition of virtually any component of a piece of equipment. Some of the technologies include

Vibration analysis

Spectrographic oil analysis

Infrared scanning

Condition-based maintenance takes predictive maintenance one step further, by performing the inspections in a "real-time" mode. This is done by taking the signals from sensors installed on the equipment and feeding the signals into the computer. The computer monitors and trends this information, allowing maintenance to be scheduled when it is needed. This eliminates error on the part of the technicians making the readings out in the field. The trending is useful for scheduling the repairs at times when production is not using the equipment.

Reliability engineering is the final step in preventive maintenance, involving engineering. If problems with equipment failures persist after using the previously mentioned tools and techniques, engineering should begin a study of the total maintenance plan to see if anything is being neglected or overlooked. If not, then a design engineering study should be undertaken to study possible modifications to the equipment to correct the problem.

Incorporating all of the preceding techniques into a comprehensive preventive maintenance program will enable a plant or facility to optimize the resources dedicated to the preventive maintenance program. Neglecting any of the preceding areas can result in a preventive maintenance program that is not cost effective.

What are the benefits of a preventive maintenance program? The most prominent benefit is the elimination of the costs related to a breakdown during production or operation. The cost to repair the equipment is easy to figure (i.e., the number of men × the hours to repair + material costs); however, this is not the total picture. The total costs for a breakdown or unexpected outage of the equipment are listed in Fig. 6-4. Most of these points are self-explanatory, but the costs may still be difficult to calculate. One of the largest intangibles is the price paid for late or poor quality deliveries made to a customer. Lost business or unhappy customers can have a dramatic impact on future business. A consulting group published figures showing in their cli-

* Operator time loss
    - Time to report failure
    - Time for maintenance to arrive
    - Time for maintenance to make repairs
    - Time required to start equipment
* Cost of repairing or replacing the failed part or component.
* Maintenance costs
    - Time to get to the equipment
    - Time to repair the equipment
    - Time to get back to dispatch area
* Lost production or sales costs or both.
* Cost of scrap due to maintenance action.

Figure 6-4. Breakdown maintenance losses.

ent base that customers lost because of poor quality can average as high as 10% of sales per year. Couple this with averages as high as 30% of the manufacturing budget going for quality problems and rework and it becomes easy to stress the importance that a preventive maintenance program can have to quality. Financial justification for preventive maintenance programs is not difficult. The costs and savings are listed in Fig. 6-5.

What type of equipment failure is it best to address with a preventive maintenance program? Since there are at least four different types of failures, an examination of them would be beneficial.

Infant mortality is a failure occurring in the first few hours of component life. This failure is understood by the electronics industry where burn-in of components are common. In this industry, the failure occurs when initial voltage is applied to a circuit and the component was not up to standard. The initial failure is called infant mortality. It is impossible to design a preventive maintenance program to prevent this type of failure.

Random failures occur without notice or warning. This type of failure is difficult to predict. This type of failure usually is engineering or materials related. These failures are unpredictable and therefore cannot have a preventive maintenance program designed to prevent them.

Abuse or misuse failures are more of a training or attitude

* Increases
  1. Maintenance personnel costs
  2. Repair parts costs
* Decreases
  1. Scrap/quality costs
  2. Downtime costs
  3. Lost sales costs

Figure 6-5. Preventive maintenance.

problem. There is no preventive maintenance program that can prevent this type of failure.

Normal wear out is the type of failure that preventive maintenance programs can be designed to prolong or prevent. These failures occur progressively over a relatively long period of time. Preventive maintenance programs can be designed to spot signs of wear and take appropriate measures to correct the situation.

How does the design of a preventive maintenance program develop? Once the decision has been made to develop a preventive maintenance program and solid management support has been obtained, the steps listed in Fig. 6-6 should be followed.

The first step is to determine the critical units in the plant that will be included in the preventive maintenance program. It has been known by maintenance managers for years that it is not cost effective to have the preventive maintenance program cover every item in the plant or facility. There are certain components, not part of critical processes, that are cheaper to let run to failure than to spend money on maintenance. Critical items should be identified and cataloged for inclusion in the preventive maintenance program.

After the units have been identified, it is necessary to break the equipment down to the component level. When they have

* Determine critical units.
* Classify units into types of components.
* Determine preventive maintenance procedures for each type of component.
* Develop detailed job plan for each of the procedures.
* Determine schedule for each of the preventive maintenance tasks.

Figure 6-6. Steps for starting a preventive maintenance program.

*** Manufacturer will provide maintenance, lubrication, and overhaul schedules.**

*** You can review breakdown frequencies and repair histories.**

*** You may consult with operators, craftsmen, or supervisors.**

Figure 6-7. Where to get information about preventive maintenance.

been broken down to this level, it is easier to develop standard preventive maintenance procedures. For example, most V-belt drives use the same maintenance procedures. Identifying all the V-belt drives makes it easier to write the procedures and apply it to all other V-belt drives, making any small changes required to customize the procedure for each piece of equipment.

In determining the procedures for each type of component, there are several sources of information that may be consulted. Several of these sources are listed in Fig. 6-7. Depending on the quality of the information, historical records can provide the most accurate information, since this information comes directly from the specific plant environment. A word of caution is required if manufacturer's information is used: these tasks and frequency tend to be too much and too often. All manufacturer's would like to see their equipment serviced as often as possible. This inflates the amount the preventive maintenance program costs, making it too expensive in some cases. Manufacturer's information is good to use for guidelines, but should not be considered hard, fast rules.

The next step after deciding on the tasks to be performed is to develop detailed procedures on how to perform each task. This detail should include the following information:

The required craft

The amount of time required for the craft

A listing of all materials required

Detailed job instructions, including safety directions

Any shutdown or downtime requirements

All of the preceding information should be detailed. For example, estimating the time required to perform the preventive maintenance, the estimate should consider all of the factors in Fig. 6-8. The more detailed the information in the plan for the preventive maintenance, the easier it is to schedule.

Preventive maintenance schedules are generally integrated into the overall maintenance schedule, unless there are personnel dedicated only to performing the preventive maintenance. In either case, the more accurate the estimates and material requirements, the more accurate the schedule. The more accurate the schedule, the more successful the preventive maintenance program. Inaccuracies lead to overscheduling, resulting in missed preventive maintenance or altered frequencies. This will ultimately result in additional breakdowns or failures, indicating that the preventive maintenance program is not successful. This results in a loss of management support, resulting in failure for the program.

One final decision needs to be made as to the types of preventive maintenance to be performed. Figure 6-9 lists the sample types of preventive maintenance. Mandatory preventive mainte-

* Preparation time.
* Travel time.
* Restrictions.
* Actual performance time.
* Area clean up.

Figure 6-8. Information that preventive maintenance estimates should include.

* Mandatory or nonmandatory.
* Pyramiding or nonpyramiding.
* Inspections or task oriented.

Figure 6-9. Types of preventive maintenance.

nance must be performed at all costs when they come due. This may involve OSHA inspections, safety inspections, EPA inspections, licence inspections, etc. Nonmandatory means inspections or service preventive maintenance that can be postponed for a short time or eliminated from the present cycle without resulting in immediate failure or performance penalty. Each preventive maintenance task should be designated in one of these categories.

Pyramiding preventive maintenance is generated each time it comes due. When there is already a preventive maintenance due and the next one comes due, the first one should be canceled, with a note written in the equipment history that the preventive maintenance was skipped. The new preventive maintenance should have a due date from the cancelled preventive maintenance written in, so that it is understood how overdue the task is. However, some companies choose to make their preventive maintenance floating or nonpyramiding. They follow the previously described scenario, except there is no notification that the preventive maintenance was missed. The previous uncompleted preventive maintenance is thrown away and the new one (without any carryover information) is issued and placed on the schedule. This results in an incomplete picture of the status of the preventive maintenance program. Instead of showing incomplete or late preventive maintenance, the system shows everything being up to date. Increased failures will result, and the preventive maintenance program showing no apparent affect on equipment operation will be looked at as worthless and lose credibility with management. This generally results in cancellation or reduction in the preventive maintenance program. To prevent this, it is essential to provide some form of a tracking mechanism to validate the preventive maintenance frequencies.

Decide whether the program is to be inspections only or task oriented. Inspections will only involve filling out a checksheet and then writing work orders to cover any problems discovered during the inspection. Task-oriented preventive maintenance allows the individual performing the preventive maintenance to take time to make minor repairs or adjustments, eliminating the need to write some work orders when turning in the inspection sheet. Of course, for scheduling purposes, there should be a time limit set on how long each task should take. Some companies set a time limit of an additional hour of work, over what the preventive maintenance was planned for. If the job takes any longer than that, the individual should come back and write a work order for someone to perform the repairs. This prevents the preventive maintenance program from accumulating labor costs that should be attributed to normal repairs.

Preventive Maintenance Program Indicators

Figure 6-10 lists indicators that show when a preventive maintenance program is ineffective. Each of these indicators can be used as an argument for improvements in an existing program or as justification for starting a program. For example, if equipment utilization is below 90%, it is obvious that the equipment is not being serviced correctly. If there is a preventive maintenance program in place, it needs rapid adjustment, before management decides it is of no value and does away with it.

If there is high wait time for the machine operators when the equipment does fail, it indicates a major failure. Major failures should be detected by a good preventive maintenance program before they occur. If there are numerous major failures, the preventive maintenance program must be changed to address the problems before support is lost for the program.

* Low equipment utilization due to unscheduled outages.
* High wait or idle time for machine operators during the outage.
* High scrap and rejects indicating a quality problem.
* Higher than normal repair costs due to neglect of proper lubrication, inspection, or service.
* Decrease in the expected life of capital investments due to inadequate maintenance.

Figure 6-10. Indicators of a poor preventive maintenance program.

As final example from Fig. 6-10, if the breakdowns can be traced to lack of lubrication or adjustments, the preventive maintenance program is to blame again. It is necessary to adjust the program quickly to address the problems. A good preventive maintenance program should remedy all lubrication and service-related failures.

* Too easy to use for "fill in" jobs.
* Failure to adhere to the schedule.
* Conflict between emergency and preventive maintenance.
* Inaccurate time and craft estimates.
* Wrong equipment being maintained.
* Insufficient detail on preventive maintenance sheets.
* Equipment failure record not available.
* Lack of monitoring and changing the program.

Figure 6-11. Reasons for failures of preventive maintenance programs.

In summary, the most common reasons for preventive maintenance programs being discontinued or being ineffective are listed in Fig. 6-11. For the most part, these problems have been discussed in this chapter. The first four items listed in Fig. 6-11 are related to the preventive maintenance programs not being given enough priority within the maintenance organization. When a preventive maintenance program is implemented, all personnel must commit to giving it the dedicated effort it will take to make it work.

The fifth through seventh items indicate that the preventive maintenance program had an improper start. It is important to make corrections in these areas as soon as a deficiency is discovered in order to keep the management support necessary to make the preventive maintenance program successful.

The last item in Fig. 6-11 indicates the need to keep the program flexible to allow for changes in the equipment during its life cycle. As the equipment ages, its requirements vary. The preventive maintenance program must reflect these needed changes if the program is to be cost effective. Failure to change or adjust will result in a preventive maintenance program that may be successful for 2 or 3 years and then become more expensive than necessary. Too much or too little maintenance with the related failures will be noted and the program will lose management support. All preventive maintenance programs must be closely monitored if they are to continue to be successful.

7 Maintenance Inventory and Purchasing

The inventory and purchasing staff have the largest impact on maintenance productivity of any other support group.

This is an interesting statement. But how does inventory and purchasing affect the maintenance organization? Figure 7-1 lists some of the ways poor inventory control can affect maintenance productivity.

In Chapter 5 the point was made that maintenance work should be planned. Part of the job plan for maintenance was the detailing of all the materials required to perform the work, ensuring that the materials were in stock and available before the work was scheduled. The list in Fig. 7-1 includes common delays in finding or transporting spare parts. If the job is planned properly, these delays will be eliminated. But if maintenance is responsible for planning the work, what do they need from inventory and purchasing to be effective? The information in Fig. 7-2 is the minimum information required.

On-line or real-time parts information is necessary to plan

1. Waiting for materials.
2. Travel time to get materials.
3. Time to transport materials.
4. Time required to identify materials.
5. Time to find substitute materials.
6. Time required to find parts in areas stores.
7. Time required to repair purchase order.
8. Time to process purchase requisitions (approvals, etc.).
9. Lost time due to
   a. other crafts without materials
   b. wrong materials planned and delivered
   d. wrong materials ordered
   d. materials out of stock

Figure 7-1. Typical material-related delays.

maintenance activities. The planner must know when selecting parts for a job that they are in stock, out of stock, in transit, etc. The planner must have current information. If the work is planned based on information that is days, weeks, or months old, when the craft technicians go to pick up the parts, they could

* On-line parts information.
* Updated stores catalogs (hard copy).
* Equipment "where used" listing.
* Part usage by cost center.
* On-hand quantities.
* Projected delivery dates.

Figure 7-2. Maintenance requirements from an inventory system.

experience all of the delays listed in Fig. 7-1. *If* the information the planner has is current, he will then know what action can be taken. For example, the minimum parts information the planner needs includes

Part number

Part description

Quantity on-hand

Location of part

Quantity reserved for other work

Quantity on order

Substitute part number

There is other information that the planner could use, but the preceding list will assist in planning most jobs. However, if the preceding information is not accurate or reliable, the planner will have to physically check the store each time work is planned. This time-consuming activity will lengthen the time necessary to plan a job properly to the point where the planner will not be able to plan all of the work required.

It is necessary to provide a current hard copy listing of all the parts carried in the stores for maintenance. This is true even if the inventory system is computerized. The catalog allows all maintenance personnel access to the stores information. This catalog is not used for planning, since the on-hand or order information would be dated. But the catalog allows maintenance personnel to know if parts are stock items or nonstock items, which can help expedite matters if a certain part is needed. The hard copy also avoids delays caused by several people looking through the storeroom to find a part that is not stocked. This situation occurs frequently during a breakdown or emergency-type repair. Providing maintenance stores catalogs at key locations can help eliminate some costly delays.

Equipment "where used" listings are lists by equipment of all of the spare parts carried in the stores. This listing is important in several ways. First of all, it allows the planner quick access to the parts information during the planning process. The planner will always know what piece of equipment the work is being performed on. The list allows a quick look-up of the part information. If the planner does not find the part on the list, this points out a possible need to add it to the list of spare parts, by requesting that stores now carry it in stock. The second situation in which this list is important is during a breakdown or emergency situation. When a part is needed, a quick scan of the spare parts list could save time looking for the part.

It is imperative the planner have accurate on-hand information. If the inventory system says there are sufficient supplies of a part in the stores to do a job, the planner may send a craft technician to get them. When the technician discovers the parts are not there, the inventory system loses credibility. This will impact the inventory system's usefulness to maintenance. If the planner or technician has to physically go to the store location and check each time a part is planned or requested, the maintenance department will experience a tremendous loss of productivity.

Projected delivery dates are important, since no store will always have every part when it is requested. Knowing when the part will be delivered allows the planner to schedule the work based on that date. This highlights the need for the delivery performance of the vendors to be good as well. There would be another loss of maintenance productivity if a job was scheduled for a certain week, only to find that when the job was started, the parts never were delivered as promised.

Until now, the points discussed have been the minimum requirements for a maintenance inventory system. Figure 7-3 lists other points for consideration that raise the level of the inventory system so that it can enhance the productivity of the maintenance department. Returns in a production inventory system indicate how many items have been returned to the vendor for some rea-

* Tracks balances for all items including issues, reserves, and returns.
* Maintains parts listings for equipment.
* Tracks item repair cost and movement history.
* Cross references spares to substitutes.
* Has the ability to reserve items for jobs.
* Has the ability to notify a requestor when items are received for a job.
* Has the ability to generate work orders to fabricate or repair an item.
* Has the ability to notify when the item reorder is needed and track the order to receipt.
* Has the ability to track requisitions, purchase orders, and special order receipts.
* Has the ability to produce performance reports such as inventory accuracy, turnover, and stock outs.

Figure 7-3. Features of a good maintenance inventory system.

son. In a maintenance inventory system, returns indicate how many parts have been planned for a job, issued to a work order, were not needed, and were returned to the stores for credit. This indicator becomes a measure of the planner's performance. This is important, since, if a planner always planned too many parts for each job, then the inventory stock level would be higher than required, which ties up unnecessary capital in spares when the company could put it to use elsewhere.

In many companies, asset tracking or movement of rebuildable spares is important. Point 3 in Fig. 7-3 shows that this information should be tracked through the stores information system.

This information is used for accounting purposes and also repair history to make repair/replace decisions. Maintenance gains very little from this information; unfortunately, many companies require this information, so tracking it through the inventory information is the easiest way to do this. Also since stores personnel maintain the parts in storage, it is easier to let them control the spares, provided maintenance can get access to the information when necessary.

Item 6 in Fig. 7-3 is also important to the planners. In many cases planners order a part for a job, holding the job until the parts arrive. Since they may plan 20 or more work orders per day, after several weeks, they may have dozens of work orders waiting on parts. It is important to have a method of notifying them when the parts are received and what work order they were reserved for. This may seem like a small detail, but it can literally save hours of work on the part of the planner.

The last item in Fig. 7-3 is also important. As with any other part of the organization, stores and purchasing should be monitored for performance. The indicators mentioned are useful to track performance levels for the stores and purchasing groups. Poor performance by these two groups will have a dramatic impact on the maintenance organization. It is good to copy maintenance managers on any inventory and purchasing reports. The maintenance manager has reference to this information for comparison with maintenance performance. Any conflicts between the two groups can then be discussed and remedied.

Organizing Maintenance Stores

Maintenance stores locations are also critical to the productivity of the maintenance personnel. Figure 7-4 lists the two types of maintenance stores options. As can be seen, these options are similar to the maintenance organizational structures. In fact, most companies that have area maintenance organizations will have

Centralized stores
reduced record keeping
reduced stores labor costs
increased maintenance travel

Area or decentralized stores
reduced maintenance travel
increased record keeping
increased stores labor costs
increased inventory levels

Figure 7-4. Maintenance stores options.

area stores locations; this situation increases the maintenance productivity by eliminating travel time to get spare parts. However, it is not necessary to have a maintenance stores location at each maintenance area shop. It is possible to locate a stores location between several maintenance areas and still have acceptable travel time to get spare parts.

Centralized stores are good for central maintenance organizations. There should be no unnecessary delays in the maintenance technicians obtaining their spare parts. It should be noted that if a central stores location is used, it should be staffed correctly so as not to create delays for people trying to get material out of the stores.

Types of Maintenance Spares

Maintenance has many different types of spares that need to be tracked through the inventory function. The list of some of the most common categories are found in Fig. 7-5. Examining these categories will help maintenance managers ensure that correct controls are placed on the more important items, while those with less importance have fewer controls.

Bin stock items are materials that have little individual value with high volume usage, examples being small bolts, nuts, wash-

* Bin stock—free issue.
* Bin stock—controlled issue.
* Critical or insurance spares.
* Rebuildable spares.
* Consumables.
* Tools and equipment.
* Residual/surplus parts.
* Scrap or useless spares.

Figure 7-5. Types of maintenance spares.

ers, or cotter pins. These items are usually placed in an open issue area. Their usage is not tracked to individual work orders as are larger items. The best way to maintain the free issue items is the two-bin method. The items are kept in an open carousel bin where the craft technicians can get what they need when they need it. When the bin becomes empty, the store clerk puts the new box in the bin, and orders two more boxes. By the time the bin is emptied, the boxes are delivered and the cycle starts over again.

Bin-stock-controlled issue items are similar to the free issue items, except that their access is limited. The stores clerk will hand the items out, while still not requiring a requisition or work order number for the item. The stock levels should be maintained similarly to the free issue stock levels using the two-box method.

Critical or insurance spares are those items that may not have much usage, but, owing to order, manufacture, and delivery times, must be kept in stock in case they are needed. The factor that must also be included in the decision is the cost of lost production or amount of downtime that will be result if the part is not stocked. It this cost is high, it will be better to stock the item than to risk the cost of a breakdown. Since the cost of these items is usually high on a per unit base, it is important that they receive proper care while in storage. This means a heated, dry, weather-proof storage area. If the spare remains in storage for 6 months, a year, or longer, good storage conditions will prevent its deterioration.

Rebuildable spares would include items like pumps, motors, gearcases, or other items that the repair cost (materials and labor) is less than the cost to rebuild it. Depending on the size of the organization, the spare may be repaired by maintenance technicians or departmental shop personnel, or be sent outside the company to a repair shop. These items are also generally high dollar spares and must be kept in good environmental conditions. Their usage, similar to the critical spares, must be closely monitored and tracked. Lost spares of this type can result in considerable financial loss.

Consumables are items that are taken from the stores and used up or thrown away after a time period. These items might include flashlight batteries, soap, oils, or greases. There usage is tracked and charged to a work order number or accounting code. Historical records may be studied and charted to determine the correct levels of stock to carry for each time. If problems develop with the stock level, the inventory level can be adjusted periodically.

In some companies, tools and equipment are kept in the stores location or in a tool crib and issued like inventory items. The difference is that the tools are brought back when the job is finished. The tool tracking system will track the tools location, who has it, what job it is being used on, and the date returned. This type of system is used only to track expensive tools or where there are only a relative few of these tools in the entire company. This system should not be used to track ordinary hand tools.

When maintenance is involved in construction or outside contractors are doing construction work in the plant, there are generally surplus or residual materials left over. Since there is no place else to put them, they end up in the maintenance stores. These residual or surplus items can become a problem in the stores. If the parts are not going to be used again in the short term (1–6 months), they should be returned to the vendor for credit. *If* they are going to be used, or are critical spares, they should be assigned a stock number and stored properly. A word of caution for these items is in order. Storing the items to have them just in case is expensive. If the store room becomes a junkyard, it is costing the company more money than most employees realize. We will examine the costs in a subsequent section.

Over a period of time, all stores accumulate scrap or other useless spare parts. At least once a year the stocking policies should be reviewed. If there are scrap items, get rid of them. One manager had an interesting method he used to clean out a stores location that was turned over to him. The manager, a supervisor, a planner, and a craft technician went through the store and identified every item they could. The rest of the items were piled outside the storeroom with a sign saying "If you recognize any of these parts, put an identification tag on them." After two weeks of people identifying parts, anything left over was scrapped. Whether managers realize it or not, it is costly to keep spares.

Another method of classifying spares is the A-B-C analysis. The outline of the system is pictured in Fig. 7-6. "A" items are high dollar, "insurance"-type items that must be in stock. It is good to have strict inventory policies on the use and movement of these items. Since there are relatively few "A" items, controlling these inventory items is not difficult.

"B" items are more numerous than "A" items, but not as costly. These items also should be controlled with a strict tracking method. By controlling the "A" and "B" items, you are controlling only about 50% of the total inventory items, but about 95% of the inventory costs. The "C" items are the open bin issue

* "A" items are
  20 % of the stock items
  80% of the inventory cost

* "B" items are
  30% of the stock items
  15% of the inventory cost

* "C" items are
  50% of the stock items
  5% of the inventory cost

Figure 7-6. A-B-C analysis.

items, which make up about 50% of the total number of the items, but only about 5% of the cost. It is a waste of time and energy to try to control the "C" items at the level you do the "A" and "B" items. The monetary return will not justify the necessary labor to process the paperwork.

One additional note on maintenance storerooms; there is a philosophy that all maintenance stores should be open. This philosophy is incorrect. As shown in the preceding discussion, it is important to have accurate and timely inventory information. There must be controls placed on the movement of certain maintenance spares. Open stores with no monitoring of the individuals having access to the stores eliminates any controls. Parts can be used without anyone knowing where they went. Someone may move them within the stores and no one else will know where they are. This type of system is expensive and will not allow a maintenance organization to use their materials effectively. Closed stores (for at least the "A" and "B" items) is critical to successfully improving maintenance stores.

The costs of maintenance inventories has been mentioned previously, but Fig. 7-7 lists some of the common hidden costs for inventory. The total cost for carrying an item in stores may be as high as 30–40% of the value of the item per year. For someone with an inventory of 10 million dollars, to think that 3 or 4

* Cost of interest for capital tied up in inventory	10–15%
* Cost of operating warehouse space, property tax, energy cost, insurance, maintenance fees	5%
* Cost of space occupancy, rent, and depreciation	8%
* Cost of inventory shrinkage, obsolescence, damage, theft, contamination	5–10%
* Inventory tax	1–2%
* Cost of labor to move in and out	5–10%
* Total of real carrying cost of the value of the item per year	30–40%

Figure 7-7. Hidden inventory costs.

million dollars are required each year to maintain that inventory is staggering. Figure 7-7 shows the critical reason why it is so important to carry only as much of each item as is really required. Anything over that amount is waste that is deducted directly from the corporate bottom line. Inventory control is critical and should not be overlooked in any effort to improve a maintenance organization.

Cost Savings Considerations

Since the costs of inventories are so high, what other efforts can be made to curb or control these costs? Figure 7-8 highlights some areas where savings have been realized in many companies. Standardization of equipment, supplies, and suppliers has proven to be a large source of savings for many organizations. For example, standardizing equipment can help reduce inven-

* Standardize plant equipment.
* Standardize supplies.
* Locate stores at key areas.
* Specify maintenance supplies and suppliers.
* Reduce or eliminate obsolete parts.
* Reduce amount of spoilage.
* Control "vanished" parts.

Figure 7-8. Areas of consideration for cost savings.

tory. Imagine a plant with 15 presses. If each press was made by a different manufacturer, how many of the parts would be interchangeable? Few, if any. What does this do to the inventory? There would have to be 15 sets of spares for each of the presses. Imagine the total cost for the inventory for such an arrangement. But what if the 15 presses were from the same manufacturer? How many of the spare parts would be interchangeable? Probably quite a few. What does this do to the inventory? Instead of 15 sets of spares, there may only be five sets. The odds of more than five of the presses needing the same part at the same time would be very small. Think of the savings in carrying costs alone for 10 sets of spares. Multiply this number times the number of different types of multiple equipment items in the plant and it can quickly add up to a very large amount.

What about maintenance supplies or suppliers? There have been studies conducted where consolidating supplies and suppliers has saved large percentages of the total inventory costs. This is one area where we can learn from the Japanese. They keep the number of suppliers low and receive better prices and service. The suppliers receive more business. The simplification of these relationships helps all involved. It is virtually an untouched area in American industries.

* Item issue quantity.
* Return policy for unused materials.
* Storage of rebuilt units.
* Order points and quantities.
* What components to stock and where to stock them.

Figure 7-9. Items that maintenance should control or influence.

Reduction of obsolete, spoiled, or vanished parts is accomplished through better inventory controls and closed storerooms, as was previously mentioned. However, these points cannot be overemphasized. There is a large savings that can be made from inventory controls and at the same time improve the service maintenance receives from the inventory and purchasing function.

Maintenance Controls

Unfortunately, there are many organizations where maintenance and stores/purchasing do not cooperate. In fact, only 50% of the organizations polled in a survey allowed maintenance any control over their inventory. This result is alarming, since maintenance is responsible for budgeting for repair materials. It is being responsible for something you cannot control. Figure 7-9 highlights minimum controls that maintenance should have over

Maintenance Serves
Operations

Stores Serves
Maintenance

Figure 7-10. Organizational order.

their own inventories. If they cannot have these controls, do not make maintenance responsible for controlling any costs, because they will not be able to do it

Unfortunately, many organizations are controlled by internal politics. Maintenance usually loses in this type of environment. Inventory and purchasing often influence upper management to a point that negates the effectiveness of the maintenance organization. Figure 7-10 highlights an important principle of organizational control. When this is pushed aside or overlooked, the entire organization suffers and frequently maintenance organizations get the blame, when it does not have control.

If the organization places emphasis in the right areas, allowing maintenance to control its own resources, maintenance can become a profit center, enhancing corporate profitability.

8 Management Reporting and Analysis

It has been said that

To manage, you must have controls.

To have controls, you must have measurement.

To have measurement, you must have information.

To have information, you must collect data.

The preceding chapters have shown where data are collected in a maintenance organization. The work order is the key document to collect maintenance information. But having the information is not important. Having the information in a usable form is important. The work order information should be used to produce reports, providing management with the information necessary to control and manage the maintenance function. Furthermore, the information should be concise and specific. Broad lists of information can be too time consuming for an manager to study.

Analysis and exception reports using the information are vital to the management of maintenance.

Beyond just maintenance's needs are the needs of the entire company. Inventory, Purchasing, Engineering, and Plant Management all need some information from maintenance. The reporting function must provide these reports as well. In examination of the reports that follow, the listing of groups that will have an interest in the reports will be included.

One final note on these reports: they are time consuming to compile manually. In organizations with any appreciably sized staff, a computer will be necessary to compile these reports. It may be output from a simple database or spreadsheet, sorted and compiled into a report. Companies using a CMMS have an advantage since many of these reports are included in the system. If they are not, then there is usually a report writer that can be used to construct these reports.

Computerized reporting has another advantage, especially with the relational databases; it can also produce meaningful graphic representations of the information. These graphs can be more helpful in describing trends and patterns than columns of figures. These graphs can also be included in reports to upper management. It would be beneficial to graph as many of these reports as possible.

Maintenance Reporting

The following reports are organized into categories with their needed review by the maintenance staff. The last general category is for reports that should be produced on an "as needed" basis.

Daily Reports

These are reports that should be produced daily for review by the appropriate maintenance personnel and managers.

Work Summary Report

This report lists each work order currently in progress or those that have been closed out in the last day. This report allows a quick look at the prior day's activity. The report should show estimated versus actual totals for the following categories:

Maintenance labor

Maintenance materials

Equipment downtime

The report should be divided into two sections—work orders completed and work orders still in process. Each of these sections of the report should be sorted by priority of the work: emergency, planned and scheduled, preventive maintenance, and routine. This summary of work performed will allow the manager quickly to answer any questions that operations may ask during a daily review meeting.

Schedule Progress Report

This report will show a listing of only the work scheduled for the week and the present status of each of the work orders. It should show the actual versus estimate figures similar to the work order summary report. The difference is that only work orders appearing on the schedule will be listed in this report. This report should conclude with a summary to show the number of work orders scheduled to be closed out during the week versus those that have been closed out. A percentage of scheduled work completed could even be included. This allows the manager to monitor schedule completion and make any adjustments necessary during the week to ensure schedule completion.

Emergency Report

This report will list all of the emergency work requested in the last day. It should be a two part report. The first part should be a line summary showing labor hours and materials dollars used. The second part of the report should be more detailed showing what craft technicians worked on the job, what parts were used, and any detail or completion notes. This allows the manager a quick look at the breakdowns/emergency work for the last day. If a particular breakdown/emergency needs clarification, he can reference the second part of the report to get the required details.

Reorder Report

This report is a list of all inventory items that have reached their reorder point during the last day. Depending on the maintenance and purchasing relationship, this list is used to generate purchase requisitions or purchase orders. If maintenance is not involved in the ordering, it may be an information only report for maintenance. If the organization is multiwarehoused, this report should be divided by warehouse. This allows transferring parts between warehouses or shows the need for parts to be ordered. The report should show on-hand, reserved, and minimum quantities for each item listed.

Meter/Predictive Work Order Report

This report will list all of the meter, predictive, or condition-based preventive maintenance (PM) work orders generated during the prior day. Depending on the sophistication of the PM system in use, these may be manually generated or automatically generated through real-time interfaces to computer systems. In either case, these work orders need quick attention, usually within a week and some may even require adjusting the present schedule to avoid a serious failure. This report should list the equipment, type of PM, type of work to be performed, and labor and material requirements.

Personnel Summary Report

This report lists all the employees that worked during the previous day, the hours they worked on each work order, and any overtime that was worked during the period. This allows the supervisor a quick review of the previous day's activity. The purpose is to ensure each employee worked and was credited for the proper number of hours on the preceding day.

Work Order Listing

This is a listing of work orders currently in the backlog. This listing should be available several different ways. Some examples could be

by requesting department

by equipment identification

by craft

by supervisor

by planner

Each of the reports should be sorted in descending order from work orders in process to work orders just requested. This allows for complete information on all work when required.

Weekly Reports

In addition to the daily reports, each week there is additional information required to manage maintenance properly. This section will provide some examples of these reports.

Schedule Compliance Report

This report compares the results from the previous week's activity to the schedule for the same week. It should begin with all work completed, all work not completed, and, finally, all work not started. This allows for detailed analysis of the week's activity. There should also be a comparison of time allotted for emergency activities versus the time actually used for emergency activities. This will give an indication of why more or less work was completed than scheduled. There could be a comparison of work scheduled to be worked on versus total work completed. This summary figure could be used as an efficiency percentage. Tracking this percentage over a 6–12 month time period will provide a complete picture on scheduling efficiency.

PM Compliance Report

This report lists the PMs scheduled to be completed for the previous week and their present status. This would be broken down by:

PMs completed as scheduled

PMs started as scheduled, but not completed

PMs scheduled, but not started

This gives a quick overview of the status of the preventive maintenance program for the previous week. A detailed section for the report may also be used to show who performed the work, what parts were used, completion comments, any related work orders that were written because of the PM, etc.

PM Due/Overdue Report

This report lists all overdue PMs. It should sort the PMs from oldest to newest. If possible, the report could start with a specified time period, like over 8 weeks overdue and sort down to the ones that became overdue last week. This allows for a quick look at the numbers critically overdue, down to those just overdue. This report would be most beneficial if a summary line of the numbers and percentage of work contained in each overdue category were listed. This report should be a good indicator of the condition of the PM program.

Schedule Projection Reports

This is a related series of reports that should be produced upon setting the next week's schedule. With the work orders identified for the schedule, the information for these reports should be pulled from the work order estimates.

Required Manpower This report should show the required craft hours for the work scheduled. The report should first list by craft a summary of total hours required. The second part of the report should be the detailed description of the work. This allows the supervisor a quick or detailed overview of his craft groups for the next week. If the schedule is produced by crew or department, then this report should be listed the same way.

Required Downtime This report lists all of the equipment downtime required for the work scheduled for next week. This report should be divided by equipment, department, line or plant, depending on the layout of the plant or facility. This is a critical report, especially for JIT, MRP, MRPII, or CIM installations. Inputting this information into the production control system is the beginning of integrating maintenance management.

Required Parts This report lists all of the parts required for the work that is scheduled for next week. This report should be given to the stores/inventory personnel to ensure the parts are ready. If the company practices staging, this is a good report for them to use to pick the materials and send to the staging area.

Required Contractors This report lists all of the contractors required for the next week's schedule. This report should list the contractors and then the work orders they are required to complete. This reduces the size of the report when there are multiple work orders for one contractor.

Required Equipment This report lists all rental, lease, or specialized in-house equipment required to complete the work on next week's schedule. This allows the planner or coordinator to ensure the equipment is ready for delivery or delivered before the work is started.

Required Tools This report is generated by companies that have tool rooms for storing general maintenance tools. It will list the required tools for the week and the work orders they will be needed for. This allows the tool room personnel to ensure all the tools required will be in good repair and ready for use.

Work Order Status Report

This report is a listing of all work orders in the backlog. It should be available in several formats.

By Department This report would be distributed to the departmental supervisors so they can see what the status is of all of the work requested for their department. This allows them to see the status of their "pet" projects without continually calling maintenance.

By Equipment This report allows quick access to all work presently requested for all equipment. This is a quick reference report for anyone requesting work, which prevents duplicate work orders in the system. They can review if a job or similar job is requested for the equipment.

By Planner/Supervisor This report gives some indication of the workload of both the planner and supervisor. Balance can be achieved in the organization by offloading work from one planner or supervisor to others during peak work times.

Past Due Work Order Report

This report is a listing of all work orders in the current backlog sorted by the "date needed" field. This report should be sorted from the most overdue to the least overdue. Work orders without any date needed information should not be included on this report. The status of all work orders should be included on this report (except completed). The reason is that if a work order is being held up for materials unavailable (for example), then, if it is overdue by a considerable amount, the planner may be able to find some manner of expediting or substituting the materials to complete the work order. This report can also provide some good feedback information to engineering, stores, and operations.

Backlog by Craft/Crew/Department

This report is used to track the amount of work for each craft or crew group. In some organizations, the work may even be backlogged by department. This should be a summary-type report, showing the totals for each craft group in total hours and then in number of weeks work. A second option could be a trending report showing the weekly backlog for each craft group for the last year. This will highlight trends, seasonal peaks, etc., allowing management to make good, justifiable staffing decisions.

Monthly Reports

In addition to looking at the daily and weekly information, there is other information that becomes more meaningful when reviewed on a monthly basis. Some of these reports are described in this section.

Completed Work Order Report

This report is a two part report (summary and detail) of all work orders completed during the month. The summary report should show total work orders closed, actuals versus estimates for the man-hours, labor costs, material costs, contractor costs. The summary should be available by several sorts:

By requesting department

By equipment type

By equipment identification

This style of summary report will assist in spotting trouble areas.

The detail report is a work order by work order listing of all the work orders. The sorts used for the summary reports should also be used for the detail report. This allows the manager the ability to investigate any discrepancies in detail.

Planning Effectiveness Report

This report is used to highlight the effectiveness of the planners. The report should show all the work orders for each planner, listing the actual costs compared to the planned costs. In some computerized systems it may be helpful to specify a tolerance percentage. This would reduce the amount of information that the manager would have to look at. A summary by planner is beneficial, with a listing of the most effective to least effective planner at the conclusion of the report. Any major discrepancies should be reviewed in detail.

Supervisory Effectiveness Report

This report is identical to the planner's report only it shows the effectiveness of the supervisors. The report should show all the work orders for each supervisor, listing the actual costs compared to the planned costs. In some computerized systems it may be helpful to specify a tolerance percentage. This would reduce the amount of information that the manager would have to look at. A summary by supervisor is beneficial, with a listing of the most effective to least effective supervisor at the conclusion of the report. Any major discrepancies should be reviewed in detail.

Downtime Report

This report is a comparison of the actual versus the estimated downtime for all closed work orders for the month. This is a critical report because it includes not just the planned and scheduled work but also the emergency/breakdown work. Rather than being a measure of planning and supervising effectiveness, this report is a measure of the effectiveness of the entire organization. As has been mentioned in previous chapters, PM programs, planning, supervising, stores, purchasing, and organizational coordination all play a part in controlling downtime. For this report, a specified tolerance could be beneficial in keeping the report short. Any major discrepancies should be investigated, with appropriate correctional measures taken.

Budget Variance Report

This report should compare the actual figures for all maintenance expenses to the budgeted figures for each category. Depending on the plant or facility budgeting procedures, they may be divided into labor, material, contractor, tools, etc. The report should provide an opportunity for the manager to correct potential problems before it is too late.

Craft Usage Summary Report (by Department and Type of Work)

This is one of the most meaningful reports that can be produced for analyzing a maintenance organization. It should summarize the work that was performed in each department (line, building, cost center, etc.) by craft. The summary should include the total resource expenditures. The second part should be a percentage listing of the total resources expended for the entire plant or facility and what percentage was used by each department (line, building, cost center, etc.). The final part of the report should show what percentage of the work done in each department (line, building, cost center, etc.) was emergency work, planned work, PM work, or other types tracked by maintenance managers.

General Information Reports

These reports are general listings of information and are kept for reference. They provide no analysis capability but form the catalog for each category of information.

Equipment Listing

This is a listing of all equipment that maintenance is responsible for. This report is found in two forms: summary and detailed. The summary is one line per item list of equipment, including things like ID, name, type, location, etc. The detailed listing is a page(s) listing per item (depending on the level of detail kept). This should include all of the information kept on the equipment, in any convenient form from a file folder to a computerized database.

Cross Reference Listing

This is a listing of all parts stocked in inventory as spares. The report may be a listing by equipment of all spares used for an equipment item. The report may also be by part number, which then lists all the equipment the part is used on. Both reports have value. The equipment listing helps the planner know what parts are carried as spares, while the parts/equipment report helps the inventory people identify items that may be of questionable value to carry as spare parts.

Parts Master List

This is a listing of all the spare parts carried in inventory. This listing may be in two forms: by part number or by part description. Both reports are necessary, since there will be times when the part number is not known. The descriptive lookup helps the craftworker or other personnel who are not familiar with the part number. The listing includes most of the information about the spare part and may take a page(s) to list this information.

Employee/Craft/Crew Listing

This is a listing of all the information on the maintenance personnel. This report may be produced by employee ID number or by a last name/first name basis. It then is a listing of all the personnel information, which may be a page(s) per employee.

Analysis/Decision Justification Reports

This series of reports is produced by management on an "as needed" basis. They are not simple lists, but use some advance statistical calculations to produce "intelligent output." These reports are used to make decisions. It should be noted that the reports do not make decisions, but the information, coupled with good management skills and insight, can help to manage maintenance successfully.

Statistics Report

This report is useful for spotting breakdown trends. It will list the downtime by equipment ID or by type of equipment, what caused it, and what effect it had on operations. One example of this report could be looking at all air conditioning units to see what the most common cause of failure was for a certain date range. Another may be plant lathes: What is the most common cause of breakdown causing delays of 10 hours or more for operations? Careful analysis of this report can lead to improvements in preventive maintenance programs or even in purchasing policies.

Resource Requirements Forecast (Open)

This report is a forecast of the known requirements for maintenance activities for a specified period. If the specified time period exceeds the work backlog, it basically becomes a PM forecast. The report will list the work in the backlog, by the oldest to the newest requests (PMs forecast to come due in the specified time period will be included). The labor, materials, contractor, etc., costs will be listed for the forecast period. This can help in short range (4–8 weeks) financial planning as well. It can also be a help to the related groups if they know what portion of their resources will be required from maintenance in the short term.

Equipment History Report

This report allows the detailing of any work performed on any specified equipment. The report should offer the user selection criteria, so the user does not have to go through 1500 work orders to find the desired information. Some of the sort areas might be:

type of work (P.M., emergency, planned, etc.)

equipment component

type of failure

type of activity (adjustment, calibration, part change, etc.)

date range

This report can be used to identify specific problems and the number of occurrences for any given equipment item.

Repair Length Report

This report is used to identify any repair, whether emergency, PM, or routine, that has a downtime greater than the amount specified at the start of the report. This enables the user to look at just the repairs that had a measurable effect on operations or facilities. It should be able to be sorted by equipment, type of equipment, department, craft, or any other field that is important to a particular organization.

Repetitive Repairs Report

This report is an analysis of the repairs that have been performed on the same equipment component when the repairs have been conducted more than the number of times the user enters. The report should list the number of times the repair occurred, how long the equipment was down each time, and what caused the equipment to fail each time. The report will be useful for analyzing repetitive failures for any equipment component, or if the report is expanded, it can show the failure rate for similar components on various equipment items. This may provide useful information to support purchasing or outage activities.

M.T.B.F./M.T.T.R. Report

This report calculated the Mean-Time-Between-Failures and the Mean-Time-To-Repair for a piece of equipment. If needed, the report should be able to break the information down to the component level. This information can be useful to set frequencies for PM programs and planning production schedules. The report should also be able to produce the M.T.B.F and M.T.T.R. information for failure causes and equipment components. These sections would help to highlight any problem component or consistent failure problem. This report is useful is planning the use of maintenance resources since the trouble area will be clearly highlighted.

Breakdown Analysis Report

This report analyzes all breakdown/emergency work requests. It can be for a specific equipment item, or for all the equipment plant wide. The following should also be sort or specification fields:

equipment type

line

shop

building

account code

failure cause

repair solution

This will enable the maintenance manager to find specific problem areas and work to correct them. By knowing the largest problems first, resource usage can be optimized.

Top Ten Report

While the previous report highlights specific problems on equipment, this report highlights the top ten equipment items in equipment downtime, labor costs, material costs, or contractor use. The report should be available by plant, department, building line, account code, or any other appropriate sort. This report highlights the specific problem equipment, while the previous report can be used to analyze the problem further.

PM Route Listing

In some plants and facilities, the PM tasks are arranged in a route for optimizing travel time on the part of the maintenance technicians. This report lists a particular route and all of the PM tasks associated with it. Analyzing this route with other routes in the same area may allow the manager to make some adjustments or changes to optimize the technician's travel time.

Overdue PM Report

This report shows all overdue PMs currently outstanding for any equipment item. The report can be sorted by individual equipment ID, department, line, building, account code, etc. The report should be used to highlight problems with missed or overdue PMs. This report coupled with the breakdown analysis report can spot breakdown problems that are related to poor PM follow-up.

Outage/Shutdown Report

This report allows the user to specify the shutdown or outage, and then list all of the work orders planned for that outage. This enables the users to have all the work orders, the requirements for the work orders, and the resources necessary to perform the work. The information can then be fed into the project management program for detailed scheduling with PERT or GHANT charts.

PM Efficiency and Compliance Report

This is a two-part report. The first part compares the planned versus the actual for completed PMs. It can be sorted by equipment, department, crew/craft, building, or individual equipment item. The comparisons should include the following areas:

Planned versus actual hours

Planned versus actual materials

Planned versus actual downtime

The user should be able to specify a percentage deviation, which allows for a specific report highlighting problem areas.

The second part of this report is the compliance report for the PM frequency. By specifying a percentage deviation allowed, the user should then get a report selected by the same parameters as the first part of the report. The report will then list the scheduled frequency for the PM versus the actual frequency of the PM. This assists the manager in examining compliance and scheduling accuracy for the PM program.

Parts Activity

The parts activity report shows the inventory activity for a given warehouse or all warehouses. It can list activity for a specific part, range of parts, or all parts. This report allows the manager to study the activity of given location for proper staffing of inventory personnel or distribution of warehouses.

Slow Moving Parts

This report is used to find those spare parts carried in inventory that are not having sufficient activity to be maintain in stock. The manager would not include critical spares in this analysis, since, by definition, they should be slow moving. But other parts not having activity in a specified time period could be marked for reduction in on-hand quantities or elimination from inventory. This report could be by warehouse or for all warehouses or for part groups or types of parts.

Issues/Returns Report

The issues/returns report shows the parts issued to a work order compared to the items returned to the stores as not needed for the work order. The report should also list any materials issued to the work order that were not planned. The report becomes a measure of the effectiveness of the planner's ability to estimate the parts required for a work order. The report should be sorted by planner. The columns for each planner could then be totaled and an effectiveness percentage calculated. This report, especially the total columns, will benefit inventory managers, as well as the maintenance manager.

Inventory over Max Report

Almost all inventory items have maximum/minimum levels set for them. This report lists any stock items whose on-hand quantities are higher than the preset maximum quantity. This will highlight any overages and then appropriate measures can be taken to reduce these quantities.

EOQ Report

The economic order quantity (EOQ) is the optimum quantity to be ordered when purchasing an item. It is beyond the scope and purpose of this text to go into an explanation of the formulas used to calculate this number. The EOQ is an effective order method for maintenance spares. This report examines the inventory and lists the EOQ for the inventory items. This can be then used to adjust this quantity in the part record.

Stock Out Report

This report will sort in descending order the inventory items that have had stock outs during the specified time period of the report. Items having multiple stock outs give an indication of a problem. These should be cross-referenced to the activity and planning reports to find and correct the problem. Stock outs can have high downtime-related costs. These should be corrected as quickly as possible.

Inventory Optimization Report

Since there are cost trade-offs in inventory, this report looks at the entire picture and recommends stock levels. It is similar to the EOQ report, except that it highlights the financial penalties associated with the decision. For example, it uses carrying costs, desired service level, downtime costs, lead times, etc., to calculate the differing costs to maintain stock levels. It should list several options for the user, highlighting the costs for more or fewer items in stock.

Reorder List

This report is a listing of all inventory items that need to be reordered. This report is often used where maintenance does not have a computerized tie-in to the purchasing department. It will list by warehouse all parts below their minimum quantity. The report can be reviewed and forward to purchasing for action.

Parts Receipt List

This report is a listing of all parts that have arrived and have been processed by receiving. The quantity received, where they have been stored, and the date should all appear on the report.

A-B-C Classification Report

The chapter on inventory described the A-B-C classification system. This report categorizes the inventory items into the appropriate classifications. This coupled with several of the activity reports can help the manager make adjustments to the inventory levels.

Price Change Report

This report lists all items that were ordered at one price but that were received at a different price. This indicates either a change in price or a mistake by the vendor. The purchasing buyer can then take appropriate action to determine the cause of the price change. Once the matter has been resolved, the inventory records can be updated with the correct price.

Parts Warranty Report

This report lists the warranty of each part in the inventory. This can be useful if high activity is shown on a part with a long warranty. Investigation may show a justification for a claim against the manufacturer for a refund. Close monitoring of the warranties can result in sizable refunds in some cases.

Restocking Report

This report is used in a multiple warehouse plant or facility. It shows items in stock at the main warehouse that need to be distributed to satellite stores locations. It lists what parts and where they need to go to. After this report is run, the reorder report needs to be run, since some items may fall below their minimums.

List Parts Issues

This is similar to the parts activity listing, except it lists what work order each part goes to. It is often listed by work order number, but more often it is by part, with each work order the part was issued to also listed.

Parts Adjustment Transaction

This report is a listing made after a physical count inventory has been taken. This report shows the difference between what was counted and what the inventory record shows was on-hand. In some computerized systems, this report is used to accept the physical count, and then lists the cost variance for each item and finally the report.

Overdue PR Report

This report lists the purchase requests (PR) that are older than a time period specified by the requestor of the report. This report insures the buyers consolidate the purchase requests into a purchase order in a timely manner. Since many maintenance activities are dependent on quick turnaround times, this is a key performance measurement report.

Overdue PO Report

This report shows all purchase orders that are past the expected delivery date. This report is important since many maintenance activities are based on these dates. When purchase orders are overdue, it is up to the buyer to call the vendor, check on the status of the order, and do what is necessary to expedite the order.

Vendor Performance Analysis

This report is an analysis of the on-time delivery, price, and total amount of business conducted with each vendor. This can be used to work for discounts, insist on quality, or other performance-related issues. This report can helpful in determining vendor of choice for purchasing critical items.

PO History Report

This report is for tracking the receipt history of a purchase order. Since many purchase orders will contain multiple line items, the parts will be received in partial shipments; this report will list each purchase order and all shipments received to date against it. This also is an indicator of the vendor performance.

Outstanding PO Report

This report is a listing of all outstanding purchase orders for any or all vendors. These may or may not be overdue purchase orders. It also complements the vendor performance report, since it is history and this report is the current outstanding purchase orders.

Cost Variance Report

This report synchronizes all of the part prices throughout the inventory/purchasing/receiving process. Any parts with a difference in price in any of these areas should appear on this report. This will ensure accurate parts costing in the maintenance work order system.

Overtime Report

This is a two-part report, with the first part being an employee listing with the total overtime they have worked for the specified reporting period. The second part of the report is a call list based on company policy to be used for contacting employees for working overtime. These reports can be time savers for supervisors in charge of contacting employees for overtime.

Cancelled PM Work

This report is a selective report by equipment ID highlighting any cancelled PMs for a specific equipment ID. This report is used to check PM compliance when a breakdown occurs. If the PM program has been neglected, this could be the reason for the failure. This report can be invaluable in justifying to upper management that PM programs are necessary.

9 World Class Maintenance Management

World Class Manufacturing—a status? or a buzzword? What is it? What does it mean? Are other countries achieving this status? Does maintenance affect an organization's ability to reach this plateau? How can a company achieve world class status for their maintenance organization? These are all good questions. This chapter will answer these and other pertinent questions.

What is World Class Manufacturing?

The term world class evolved from a situation that has developed in the marketplace over the last two decades. In the 1960s and early 1970s, we had both domestic and international markets. If a company chose to compete in the domestic market, it found the competition was from within the United States as well. If they had a shortage of raw materials, a trucking strike, a rail strike, or some other interference, their competitors suffered the same problem. All other things being equal, it often came down to price, quality, and delivery. Many times there was little difference between the vendors in the domestic markets.

If U.S. manufacturers chose to enter the international markets, they found they had superior products, technology, and marketing skills. They could control most of the international market. To give an indicator of this, the U.S. in 1965 had a 30% share of the world's market of manufactured goods. A country with less than 1/10th of the world's population, producing 1/3rd of the goods manufactured. In the late 1960s and early 1970s, the "fat cat" attitude developed. American industry could do no wrong. They could sell whatever was made, no matter what the cost, no matter what the quality.

But while the United States was in this dream world condition, other countries became envious. Governments, in an effort to spur their economies, encouraged economic development. These countries began to change their attitude. They analyzed what the United States was doing, and decided to do the same. As they worked 16–20 hours a day retooling, rebuilding, developing their economic potential, a strange thing happened. Just as a runner who has trained and surprises both himself and an overconfident competitor who has not trained because he saw no challenger in the field, the market reversed. Now the Europeans and the Asian countries were making inroads, not only in their own market, but in the U.S. markets as well. The U.S. companies were shocked into indecisiveness. The recession of the mid 1970s was reducing the U.S.'s ability to recover and reinvest. They began sacrificing long-term gains for short-term profits to survive.

The late 1970s and early 1980s was a boom time for the foreign competitors. Their new found advantage had turned into an opportunity to floor U.S. companies. They continued to increase quality, to lower prices, and to deliver products just in time. They developed a competitive edge that would be difficult to overcome. The demand for foreign goods in the United States would continue to increase until, in 1987, the United States showed a deficit of 170 billion dollars in manufactured goods. A comparison of trade balances is depicted in Fig. 9-1. The magic formula of price, quality, and on-time delivery and all other related factors were working for the foreign competitors. As Fig. 9-2 indicated they had succeeded in reducing the U.S. market share to 10% of the worldwide market in 1987. A 20% decrease in just over 20 years; was the giant dead?

The true meaning of "World Class Manufacturing" now becomes clear; the ability to compete anywhere in the world, to be able to meet and beat any competitor anywhere in the world with product price, quality, and on-time delivery. How to accomplish this is the real problem. However, the answer can be divided into three areas: quality, attitude toward competition, and automation technology.

Quality

In a paper presented at the AIPE national conference, it was pointed out that 20–25% of the average manufacturing operating budget goes for finding and fixing mistakes. If the cost of repairing or replacing flawed products are included, this figure may be

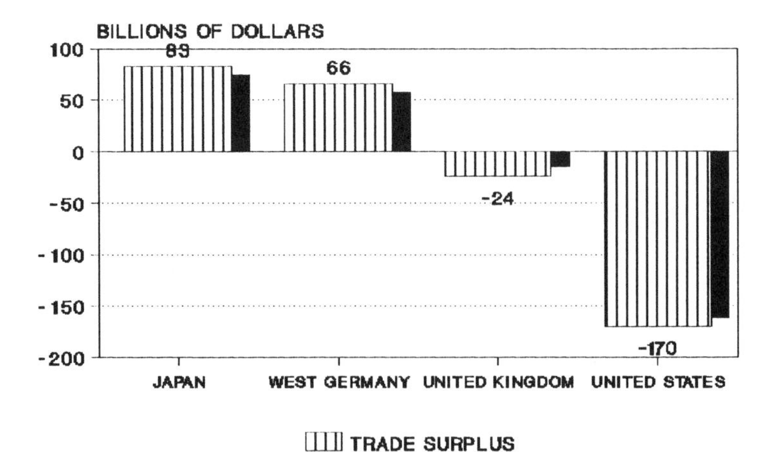

Figure 9-1. Who is winning the world class competition?

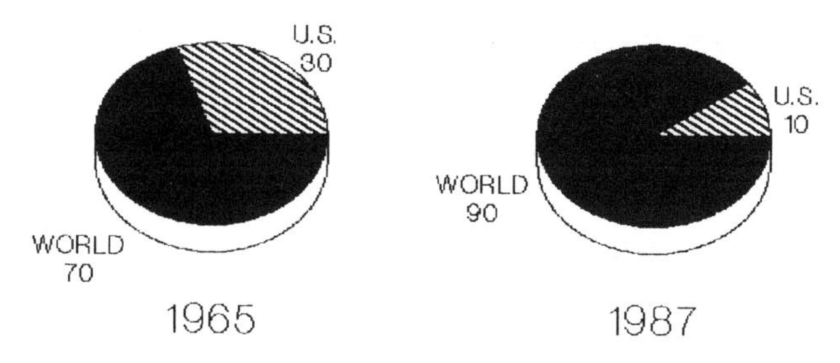

Figure 9-2. Market share of manufactured goods.

as high as 30%. Who could calculate the cost if it included the lost future sales due to a poor reputation for quality? One consulting company figures that the cost for poor quality is as high as 10% of sales per year.

The U.S. companies have made inroads in this area; however, the made-in-the-U.S. label has not regained its appeal, when we look at Japanese quality. The entire concept of the Japanese is to prevent defects, not to correct defects, which is the common attitude in the United States.

Attitude Toward Competition

This is one of the most difficult attitudes to adjust. Moving to a world market, the perspective changes. We cannot sit back and wait for the government to impose restrictions and save any industry, no matter how fundamental it is to the U.S. economy. If a U.S. company is to remain in business, the answer lies in internal improvements. We must view each competitive situation as serious. The thought that "the customer will order from us because we are an American company" is fiction. They will buy where the price, the quality, and the service are the best.

Automation Technology

In an article published in *Plant Engineering Magazine* (3/12/78) it was stated that only 20% of the plants in the United States are automated at the level of our overseas competitors. This is a large advantage that U.S. companies are giving their competitors. If these companies are going to lower costs and raise quality, automation is the solution. The Japanese and the Europeans have many industries that are totally automated from order entry to delivery, from product design to product production.

A warning to U.S. companies, however, is that while automation is the answer, automating waste is not. World class manufacturing requires the elimination of complexity. It requires simplicity in design and manufacturing processes. Simplicity is the key to eliminating waste.

How Does Maintenance affect an Organization's Ability to Achieve World Class Status?

Since it is true that achieving world class status involves quality, attitudes, and automation, it would be wise to examine how maintenance is involved in each of these areas.

Quality

If the equipment is operated in an environment where no maintenance is performed if the equipment is running, it will be in poor shape. All equipment requires maintenance activities, whose frequency varies with the equipment's age. If the equipment does not receive the proper maintenance at the proper interval, it cannot maintain standards. Poorly maintained equipment seldom produces consistently good products. If quality programs are instituted in a company, maintenance must be a part of it.

As previously discussed in Chapter 6, there are several parts to the program. Inspections of equipment in a world class organization may be performed by the equipment operator. If properly documented, these inspections can help to trigger replacements of defective components or necessary adjustments. This relieves some of the less technical work from the maintenance organization. This also helps to promote good relations between maintenance and operation. It also will help to increase the quality of the product.

Predictive and condition-based maintenance systems will help to ensure higher quality products, since the equipment condition should never deteriorate to the point where it will be producing a defective product. Changing the present "run-all-you-can" or "run-till-it-breaks" mindsets of plant managers will involve the next section.

Attitudes

One of the prime attitudes here is the sacrificing of long-term planning for short-term gains. While this has one meaning in the world market and the product a company plans on putting there, it has a meaning for maintenance as well. If a company management has this attitude, it will affect the maintenance organization in the areas of

Preventive maintenance

Labor planning

Inventory

Preventive maintenance suffers, since no one is willing to give the equipment to the maintenance department for routine service. Since this may involve some lost production, it is necessary to view matters over the long term. Keeping the capital equipment in good repair, producing high-quality products, should be the goal of maintenance and operations. If they cooperate, long-term savings and higher product quality will result. For example, one company after instituting a preventive/predictive maintenance system estimated its savings to be as high as 1.2% of the total plant output. However, it should be noted that the costs for the PM program start-up pushed maintenance costs up for 6 months before results were shown. Short-term versus long-term savings; it is an attitude that must be adjusted.

Another attitude that must be adjusted is that of management toward the position of maintenance in the organization. By far the organizations that have been successful in raising their maintenance organizations to a world class standard have had a management group that was committed to the goal of putting maintenance management on the level as other management positions in the organization. Without this type of commitment, maintenance cannot improve. The foreign competitors understood this some time ago. Their organizations hold maintenance people in respect for the contributions they can make to overall profitability.

Management's attitude toward maintenance resources must also change. What resources? Maintenance resources are

1. Labor
2. Materials
3. Tools
4. Supplies and miscellaneous

Examine the maintenance budget for your organization. How much is spent annually on maintenance salaries? How much is spent on maintenance parts? These figures can be astounding. But for a moment, consider these points: When was the last time you heard of a company begin production on a product without considering

the material required

the equipment required

the labor required

the time required to produce the finished product

This would be ridiculous to even consider, wouldn't it?

Could you imagine: In a world class environment

* Each operator waiting for someone to tell him what equipment to operate? Or telling management what product they are going to make today?
* Each operator going to the stores to get the material needed to make the product as his process required it?
* Each operator ordering their own material to make their assigned product and then waiting on it?
* Each operator waiting in line to use the equipment to make their assigned product because others needed the same equipment at the same time?
* Operators standing, watching another operator work, because no one knew it only took one person to run the machine, and scheduled too many people?

Figure 9-3

Even more ridiculous is the scenario in Fig. 9-3. After considering Fig. 9-3, would you want to own such a company? Would you like to invest in such a company? Would you like to work for such a company? Your truthful answer would be *NO* to all three questions, wouldn't it? Why would you so answer? Because you realize that any company operating in this type of mode, would never be profitable, would be extremely wasteful, and would be frustrating to work for.

If we would feel this way about a company that operated in this manner, why don't we feel that way about our maintenance organizations? Most organizations operate similar to the scenario painted, with varying degrees of difficulty. Why do they do this? It is because they are forced to by shortsighted management, who place no importance on maintenance. Unless there is a change of attitude, these types of maintenance organizations have little or no chance of improving. *MAINTENANCE PLANNING IS THE*

MOST ESSENTIAL PART OF ANY COMPANY'S EFFORT TO IMPROVE PERMANENTLY THE MAINTENANCE ORGANIZATION.

In summary of this topic, management must understand maintenance management if it is to improve. If they do, they will reap the following benefits:

Quality improvements

Improvements in utilization and uptime

Reduction of maintenance labor and material costs

They will also have an organization that is prepared to do business in the future world market, no matter how severe the competition is.

Automation Technology

What does automation technology mean to the maintenance organization? It means being able to take advantage of all of the technological advances in maintenance equipment to keep maintenance costs at an optimum level.

Examples of automation technology in maintenance might start with vibration analysis, which started as an engineering function almost a decade ago. As the technology matured, the tools became easier to use. Soon, it was common for maintenance technicians to be involved in making and trending the readings. Complementary software was soon developed that made the charting, trending, and interpretation of the data even easier. This situation continued to develop until it is the exception to find an organization today not involved in some form of vibration analysis. It would be difficult to calculate the savings within companies that this predictive technique alone has saved companies.

Other areas of automation technology could include additional predictive maintenance techniques such as infrared scanning, or spectrographic wear particle analysis. While these techniques are just now coming into wide use, they have the same potential as vibration analysis to improve maintenance performance and utilization.

Another area just being developed is condition-based maintenance, in which real time inputs are fed to maintenance where the parameters can be analyzed and appropriate action can be taken. It will be several years before this technology becomes widely accepted.

The computerization of maintenance information is another example of automation technology. The common term for this type of system is a Computerized Maintenance Management System (CMMS). These systems automate the paper flow used by the entire maintenance organization. These systems consist of the following subsystems:

Equipment data

Preventive Maintenance

Work orders

Inventory

Purchasing

Personnel

Reporting

The advantage of using a computer to automate the maintenance information is that it allows management access to historical data, concise summary reports, with the graphic display of the data. However, some companies have gotten buried in the amount of data they can produce with one of these systems. The information gathered should support management, not burden it.

CMMS appears to be the final key to allowing maintenance to take full advantage of automation technology. Vendors, consultants, and software experts are working to pull information from predictive maintenance systems, computer-based training systems, and other related packages into CMMS. One example includes using the predictive maintenance information to trigger repair work orders in CMMS, where they are planned and scheduled according to their priority.

The technology will continue to advance and will be used to optimize further maintenance resources. World class organizations will take advantage of these advances. They will have to, since their competitors will be doing so. Procrastinators will lose out, because the winners are making their moves now. The question becomes: Will you move your organization into the world class arena?

10 Integration of Maintenance Management

CIM (Computer-Integrated Manufacturing), MRP (Material Requirements Planning), MRP II (Manufacturing Resources Planning), JIT (Just-in-Time), FMS (Flexible Manufacturing Systems), Shop Floor Control, CAD/CAM (Computer-Aided Design/Computer-Aided Manufacturing) are all acronyms that are becoming commonplace in manufacturing today. But, yet, as much of the literature on these topics is analyzed, there is almost no mention of the maintenance organization. There is no mention of the interaction maintenance management is going to have to have with other parts of the organization to make these systems successful.

The problem is one mentioned earlier—attitude. Engineers and managers who conceive and implement these systems are forgetting one fact: to manufacture the product you need equipment. To operate in a world class manufacturing organization, you need automated equipment. To operate automated equipment, you need automated control systems. To ensure the automated control systems and the automated systems function, you need a high-level or world class maintenance organization. If maintenance is not included in a business plan for automated manufacturing (not matter what acronym you choose to use), it will not be successful in the long term.

While the last statement may seem to be rather bold, it is true. There will be some that point to a factory, and show where maintenance was not involved. The factory is operating relatively maintenance free. In the light of the last paragraph, how can this be? It is because all of the equipment is new. This means that it has not aged to the point where maintenance is required. Depending on the process and loads put on the equipment, see how the condition deteriorates in 3–5 years. The breakdown rates climb, while the utilization rates drop. Some companies decrease the yield of the equipment to compensate for the amount of maintenance-related equipment downtime.

No matter how companies try to cover it over, maintenance is an important part of any world class manufacturing organization. How does maintenance fit into the total picture? We will now examine an example.

Maintenance and MRP

MRP systems use the master production schedule and the bill of materials to determine what equipment will be required, what parts are necessary, and what labor is needed to produce a product. The master production schedule has all of the sales forecasts, allowing the output for the company to be known. The bill of materials also helps to determine what the total materials demands will be from stores. The production scheduling knows how long each item takes to produce on each piece of equipment, allowing the labor resources to be determined.

This process seems simple enough and can be very accurate. But inaccuracy in the system develops when the assumption is made that the equipment will be available when necessary to produce the product. If the schedule requires the equipment to run for 16 hours a day for 5 days during the week, the equipment must run at standard operating speeds, producing a quality product for the 80 hours.

The equipment breaks down stopping production for 4 hours on three separate shifts. This lost 12 hours plus the 8 hours where the hydraulic system would not develop enough pressure to allow normal operating speeds and the equipment had to run at 50% speed, would total 16 hours of lost production. How would this lost production be made up? There would be two options, run two extra shifts on the sixth day, or push the added demand into next week's schedule. Pushing the orders back into the next week is not an acceptable solution to the JIT environment. This would delay the order, possibly not allowing it to arrive at the customer's site in time for their needs. This can result in the loss of a customer in today's competitive marketplace.

The only acceptable solution is to operate the two extra shifts. This enables the production to be made up with no appreciable delay. But what are the costs involved? Here is a partial list:

Overtime production labor

Overtime maintenance labor

Extra utilities to run equipment

Extra time to redo the production schedule

Extra time to inform customers that their orders may be delayed

The costs for these extra shifts could have been avoided with a preventive maintenance program. In fact, most "real world" experts agree that the only way most of the "new" production methods will work is with a rigid preventive maintenance program. In considering the previous example, what if a preventive maintenance program had been in place?

If the preventive maintenance program included some of the newer technologies, like vibration analysis, it is probable that the causes of the breakdown would have been detected before the breakdown occurred. The repairs could have been scheduled during the off shift through the week, eliminating the delay for the production unit. The clogged filter, reducing flow on the hydraulic system, with the subsequent loss of pressure, would have been replaced on an off-shift during the routine PM service for the hydraulic system, thus eliminating the lost production time due to reduced speed operation.

While these may seem like simplistic examples, they have actually occurred at some sites. Other examples could have been used, for every company has similar stories to tell. All include unnecessary breakdowns that could have been avoided with a management structure that gives importance to maintenance management. How does the maintenance organization fit in the automated factory, the JIT environment? There are three options: basic, interfaced, and integrated.

A *basic* solution is the manual input from a maintenance program into the production scheduling process. This may involve the maintenance manager comparing schedules with the production manager to see if there is any conflict between the time maintenance requires the equipment for service and the time operations needs to make product. The conflicts can be resolved on a case-by-case level, allowing the decision to be made to risk a problem or to correct it. The plant manager should resolve any decision the maintenance and operations managers cannot.

An *interfaced* system is used when both the production planning process and the maintenance management function are computerized. This allows the two systems to operate independently of each other except when information is batch loaded from one to another. The most common example is loading the maintenance demands for time to the production scheduling part of the production planning system. The production planning system would treat the maintenance demands for the equipment just as if they were product demands. This would allow for a smooth scheduling process. If there are no conflicts, the schedule could be produced and finalized. If there are conflicts, they could be handled by shifting resources, off-loading production to other equipment, or postponing the maintenance request (if it is not urgent) to the next week. This allows the two systems to work together, avoiding the "islands of automation" problem.

The most advanced system, which is just now beginning to be used, is an *integrated* system. This is different from the interfaced system because it is real time, not batch loaded. This type of system will become even more important when the PLCs on the shop floor are feeding back information into the production scheduling and maintenance systems in a "real-time" environment. This is the condition-based environment that maintenance organizations will have to move to in the future if they are to contribute to corporate profitability in a world class organization. In this organization, all production information is fed back into the production scheduling system for compliance checks against the master production schedule. Part of the same information (run times, production rates, etc.) is also fed into the maintenance system. This information coupled with the data from vibration, temperature, or sonic sensors is used to schedule maintenance just-in-time to prevent equipment outages or quality problems.

This synergistic relationship will be necessary for the company to achieve optimum cost/product/service relationships. Only through achieving this type of cooperation will any company be able to remain competitive in the world market. If companies delay in adding maintenance to their total plan for organization improvement, the world market will soon pass them by.

Maintenance has been called "the last frontier" for management to conquer. Just the same as there have been pioneers in the past who have gone out and explored the frontier ahead of the crowd, so there have been some pioneers in this area as well. They have been able to take advantage of the cost benefits available. You are able to read of their exploits in magazine articles or listen to them at conference presentations.

Just the same as pioneers would come back and tell their tales, spurring others on to move into the frontier, so it is hoped that others will begin to move into the last frontier for manufacturing and facilities. There will be problems, but as has been the American heritage, overcoming the problems brings a sense of pride and accomplishment. The final question for this book is "Are you and your organization up to the challenge?" Innovators are making things happen. Procrastinators will not conquer the final frontier. The choice of which group you belong in is up to you and your organization. The choice will have lasting effects on the your survivability. It is hoped the information in this book will help you make the correct decision.

Index

apprentice training, 56

breakdown, costs of, 105
breakdown maintenance, 39–40

combined in-house/contract maintenance staff, 42
complete in-house maintenance staff, 41–42
Computerized Maintenance Management System (CMMS), 161–162
condition-based maintenance, 40–41, 104, 161
contract maintenance staff, 42
  complete, 42–43
craft backlog, 92, 93
craft training programs, 56–59
cross training, 60–62

delays, material related, 115

emergency breakdown work order, 70–71
equipment service level, 39–41

hands-on time, definition, 89

inventory costs
  cost saving considerations, 125–127
  hidden, 124–125

journeyman training, 59–60

maintenance controls, 127–128
maintenance inventory and purchasing, 114–128
  features of good, 118
  requirements of, 115

maintenance labor capacity, 91
maintenance organizations
  analyzing, 12–31
  attitude of management toward, 52
  geographical, 43–45
    area, 43–44
    centralized, 43
    combined, 44–45
  goals and objectives, 32–38
  growing, 45–47
maintenance planners, 76–77
  job description, 81–86
maintenance planning
  benefits of, 88–90
  failures in, 86–88
maintenance scheduling, 90–97, 109
  priorities, 94, 95
  work order considerations, 94
maintenance spares, types of, 120–125
maintenance staffing, 48–51, 92
  options, 41–43
maintenance stores, organizing, 119–120
maintenance supervisors, 78–81
maintenance training, 55–64
maintenance, types of, 39, 102–111
  and MRP, 164–168
management reporting and analysis, 129–151
  analysis/decision justification, types of, 141–151
  daily reports, types of, 130–133
  general information reports, types of, 140–141
  monthly reports, types of, 137–139
  weekly reports, types of, 133–137
minor lube programs, 40

planned and scheduled work orders, 69
planner training, 62–63
preventive/predictive maintenance, 33, 40, 98–113
  benefits of, 105–106
  definition of, 98
  importance of, 99
  indicators of failure, 111–113
  and JIT, 99–100
  obtaining information about, 108
  starting, 107
  types of, 102–111

shutdown/outage work orders, 72
standing/blanket work orders, 69–70
supervisor training, 63–64

"where used" listings, 117
work orders
  definition, 65
  objectives of, 69
  scheduling considerations for, 94
  status codes for, 91
  types of, 69–72
work order systems, 65–75
  obstacles to effective utilization of, 72–74